Multispectral Image Analysis Using the Object-Oriented Paradigm

Multispectral Image Analysis Using the Object-Oriented Paradigm

Kumar Navulur

CRC Press
Taylor & Francis Group
Boca Raton London New York

CRC Press is an imprint of the
Taylor & Francis Group, an informa business

CRC Press
Taylor & Francis Group
6000 Broken Sound Parkway NW, Suite 300
Boca Raton, FL 33487-2742

© 2007 by Taylor & Francis Group, LLC
CRC Press is an imprint of Taylor & Francis Group, an Informa business

First issued in paperback 2019

No claim to original U.S. Government works

ISBN-13: 978-0-367-44624-6 (pbk)
ISBN-13: 978-1-4200-4306-8 (hbk)

**Visit the Taylor & Francis Web site at
http://www.taylorandfrancis.com**

**and the CRC Press Web site at
http://www.crcpress.com**

Library of Congress Cataloging-in-Publication Data

Navulur, Kumar.
    Multispectral image analysis using the object-oriented paradigm / Kumar Navulur.
        p. cm.
        ISBN-13: 978-1-4200-4306-8 (alk. paper)
        ISBN-10: 1-4200-4306-4 (alk. paper)
        1. Image analysis. 2. Remote sensing. 3. Object-oriented methods (Computer science) I. Title.

TA1637.N37 2006
621.36'7--dc22   2006021814

Table of Contents

Preface

This book is intended for students, research scientists, and professionals in the remote sensing industry who have a basic understanding of remote sensing principles, image processing, and applications of remotely sensed data. This book will appeal to users who apply imagery to a variety of mapping applications, including vegetation mapping, identification of man-made structures from imagery, mapping of urban growth patterns, and other applications. I wrote this book keeping in mind users with diverse educational backgrounds. This book is designed to explore the new object-oriented paradigm of image analysis. The content is tailored to both novice users as well as to advanced users who are ready to use or have already tried using objects for image analysis. I cover a variety of applications and demonstrate the advantages of object-based image analysis, using step-by-step examples and, wherever necessary, explaining the functionality in eCognition to accomplish tasks.

Object-based image analysis is a paradigm shift as compared to traditional pixel-based image analysis approaches and brings a fresh, new perspective to the remote sensing discipline. Advantages of this approach are demonstrated using various hands-on exercises in this book. For example, it is efficient to identify an agriculture crop based on vegetation boundaries rather than every pixel within the field. The spatial and spectral variations within a field, due to vegetation density, crop stress, presence of irrigation, and ditches that run through the field, make classification of a crop challenging on a pixel-by-pixel basis. The new object-oriented paradigm creates objects by grouping pixels with similar spectral and spatial characteristics and reduces crop classification problems to a few objects rather than thousands of pixels. This approach lets you take advantage of the true power of remote sensing by exploiting all the dimensions of remote sensing, including spectral, spatial, contextual, morphological, and temporal aspects for information extraction.

I will demonstrate a simple example of classifying an island by using contextual information that land surrounded 100% by water can be identified as an island. Such contextual relationships are often explored in the geographic information system (GIS) field rather than in the image analysis domain. Further, morphological information such as size of water bodies can be used to discriminate between ponds and lakes. Long, narrow water bodies can be classified as streams and rivers, to which other shape-related properties such as length, width, and compactness of objects can be added. The object-oriented paradigm combines the power of image analysis with some of the GIS-analysis capabilities to empower you with new tools to extract information from remotely sensed data.

Further, this technology allows you to extract features from the same dataset with different object sizes and spatial resolutions. Often within a scene there are large features such as tracts of bare soil, forested areas, and water bodies that can be extracted at a much lower resolution as compared to more comprehensive urban features. Also, ancillary data such as elevation, existing thematic maps, parcel layers, and other geospatial information can be readily integrated into object-oriented image analysis. In this book detailed examples show you how to exploit all aspects of remote sensing step by step and the advantages of this approach are discussed.

Recent years have seen increased availability of imagery from multiple data sources, including aerial and satellites. Sample data from a variety of data sources and at differing spatial resolutions are used in this book to demonstrate the applicability of the new paradigm on any multispectral data. As we step through various examples, we will review briefly the current image-analysis approaches and demonstrate how to take advantage of these techniques to improve thematic classification results. I recommend reading the articles in the bibliography section for an in-depth understanding of the techniques.

I have dedicated a chapter to accuracy assessment of the classification results. We will review the accuracy assessment techniques currently followed in the industry and demonstrate how to assess thematic accuracy within eCognition software.

I would like to acknowledge the support of my family and friends who helped me during the course of writing this book. Special thanks to Mike McGill for encouraging me to take on this book. I would like to acknowledge DigitalGlobe GeoEye, and the St. John's (Florida) Water Management District for providing the images used in this book.

Kumar Navulur

List of Figures

List of Tables

1 Introduction

1.1 BACKGROUND

Image analysis of remotely sensed data is the science behind extracting information from the pixels within a scene or an image. Multispectral sensors for remote sensing are designed to capture the reflected energy from various objects on the ground in the visible and the infrared wavelengths of the electromagnetic (EM) spectrum of the sun. Some of the sensor ranges extend all the way into the thermal spectral range, whereas most of the commercial sensors today primarily capture data in the visible and near-infrared regions of the EM spectrum. These digital sensors measure the reflected light from various objects and convert the data into digital numbers (DNs) that comprise an image. For the past few decades, the aerial industry has used photographic film with various spectral filters to capture the data. The aerial industry is converting to digital cameras. The negatives were then scanned to create digital images of the data. In this book, we will perform image analysis using digital data acquired from both aerial as well as satellite sensors.

Traditional image analysis techniques are pixel-based techniques and explore the spectral differences of various features to extract the thematic information of interest to the end user. Although there are few instances where objects that are the size of a pixel need to be identified, typical applications involve extraction of features that are comprised of multiple pixels such as roads, buildings, crops, and others. These applications require classification of groups of contiguous pixels that make up a feature. This book will explore the object paradigm where we group pixels of similar spectral and spatial response, based on predefined criteria, and apply the traditional pixel-based image analysis techniques on objects to extract features of interest.

1.2 OBJECTS AND HUMAN INTERPRETATION PROCESS

Objects are the primitives that form a scene, photograph, or an image. For example, when we look at a photograph, we analyze it by breaking it down into various objects and use properties such as the shape, texture, color, context, and others for understanding the scene. The human brain has the innate capability to interpret the rich information content available in the scene and can intuitively identify objects such as cars, houses, golf courses, and other features present within the scene. Further, our cognizance powers allow us to instantaneously exploit the knowledge already stored in our brains and convert it into an expert rule base for image interpretation. The following paragraph demonstrates this object-oriented concept:

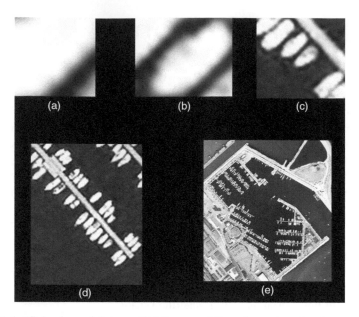

FIGURE 1.1 (Color figure follows p. 82.) Representation of a marina at various zoom levels.

Figure 1.1 shows a natural-color QuickBird scene of a marina in Italy at various zoom levels. Figure 1.1(a) is zoomed all the way to the resolution of a few pixels. At this resolution, we can interpret that it is a bright feature/pixel, but we cannot assign a feature to the object. Figure 1.1(b) is zoomed out and, now, we recognize it as an elliptical object that is bright. Now we have a starting point, and our brain can start retrieving various features that are elliptical in shape and have bright spectral reflectance and also use the contextual information that the bright object is surrounded by dark features. At this zoom level we can correlate with features, such as a whale in the water or a submarine, but it is difficult to interpret what these features are. Figure 1.1(c) shows the same scene zoomed out to a few hundred pixels, now revealing similar elliptical objects surrounded by dark features in proximity to a long, narrow rectangular object. This zoom level gives the first clue that the objects are probably boats in water. Our intuitive process puts the various pieces of the puzzle together, including various bright objects that can be boats surrounded by a dark water feature next to a rectangular pier. Figure 1.1(d) and Figure 1.1(e) reveal the other surrounding features, such as piers and docks, that our brain interprets to be a marina. An object-oriented approach is a first step in replicating this human interpretation process.

1.2.1 Human Interpretation Process versus Computer-Aided Image Analysis

The human brain is an efficient learning, storage, and quick retrieval mechanism that can store various patterns and can instantaneously retrieve the patterns from our memory banks. Also, we can identify various instances of the same object in different orientations, colors, sizes, and textures. We have the complex reasoning capability to connect the missing dots in the features such as roads occluded by trees and shadows.

Computer-aided image analysis is faced with the daunting task of mimicking the human interpretation process and associated complex interpretation rules, advanced pattern recognition, and knowledge retrieval mechanisms. Numerous remote-sensing techniques have been developed for analyzing the images, such as supervised classification, texture analysis, contrast enhancement, and others that exploit the spectral responses of various features in different multispectral bands. Significant progress has been made in the field of artificial intelligence (AI), fuzzy logic classification, rule-based classification from the expert systems domain, and neural nets in capturing human ways of thinking and applying them to image analysis. Until recently, most of these techniques were applied to pixels, limiting the use of these advanced technologies. This book will demonstrate how the new paradigm of an object-oriented approach can take advantage of these technologies and apply them to improve the image analysis and interpretation process.

1.3 OBJECT-ORIENTED PARADIGM

An object can be defined as a grouping of pixels of similar spectral and spatial properties. Thus, applying the object-oriented paradigm to image analysis refers to analyzing the image in object space rather than in pixel space, and objects can be used as the primitives for image classification rather than pixels. Image segmentation is the primary technique that is used to convert a scene or image into multiple objects. An object has, as compared to a pixel, in addition to spectral values, numerous other attributes, including shape, texture, and morphology, that can be used in image analysis. Figure 1.2 shows objects created for the scene with the marina in Figure 1.1.

As I will demonstrate in later chapters, for the same scene, objects can also be created at different sizes and at multiple levels from the same image. The advantage of such an approach is that you can mask out objects of no interest at larger scales, and focus on extraction of features of interest to the end user. Further, by creating objects at different levels, parent–child relationships can be leveraged to improve/enhance feature extraction process in applications such as change detection.

The object-oriented paradigm will allow us to exploit all aspects of remote sensing, including spectral, spatial, contextual, textural, and temporal properties for feature extraction. In this book I will introduce you to this paradigm of image analysis and will demonstrate the power of the new approach by taking you step by step through various hands-on examples.

1.4 ORGANIZATION OF THE BOOK

The material in this book is organized in six chapters, divided into three broad topic areas: (1) traditional image analysis techniques, (2) basics of object-oriented technology, and (3) application of image analysis techniques using the new paradigm.

Chapter 2 will have a brief overview of the traditional image analysis techniques.

FIGURE 1.2 (Color figure follows p. 82.) Segmented image of a marina.

Chapter 3 will introduce the basics of object-oriented technology and discuss some of the benefits of object-based classification versus pixel-based classification, using some examples. I will provide an overview of various object properties that are available for image classification.

Chapter 4 will show the creation of object primitives using image segmentation approaches. We will also cover the basics of eCognition software, including creation of objects, perform basic image classification using objects, and create multiple levels for information extraction. We will do an end-to-end supervised classification using an object-oriented approach by picking training samples and also conduct a thematic accuracy assessment of the final classification.

Chapter 5 will briefly discuss various image analysis techniques, and using exercises, demonstrate the use of these techniques for object classification. We will cover the basics of the AI techniques used in image analysis. We will demonstrate by examples how to make use of the spectral, spatial, contextual, and temporal dimensions.

Chapter 6 is dedicated to advanced image analysis. I will discuss various image enhancement techniques, how to use ancillary data to constrain image segmentation, demonstrate the concepts of semantic grouping of objects, and show examples on how to take advantage of these advanced concepts in eCognition.

I discuss accuracy assessment approaches in Chapter 7 and demonstrate the functionality within eCognition to assess accuracy.

2 Multispectral Remote Sensing

In this chapter we will focus on various aspects of remote sensing that will help you gain the understanding of the basics for choosing the right data set for a given application. Remotely sensed data available today is captured using multispectral sensors mounted on aerial as well as satellite platforms. Imaging using aerial platforms/aerial photography has been around for several decades and has been primarily using film as the medium to image the earth. Recent years have seen the conversion from film-based sensors to digital sensors, and the proliferation of digital aerial sensors is resulting in more access to very high-resolution remotely sensed data sets across the globe. The launch of the Landsat program in the early 1970s signaled the beginning of the satellite remote-sensing era. Until the late 1990s, medium-resolution data sets were the common source of satellite imagery for the remote-sensing community. The launch of the IKONOS satellite in the late 1990s with 1-m pixel resolution signaled the advent of high-resolution satellite imagery followed shortly thereafter with SPOT 5, IRS, QuickBird, OrbView3, and other high-resolution satellites. Approximately 10 to 15 high-resolution satellites are scheduled to be launched in the next few years allowing the end user multiple choices of data sources as well as different spatial resolutions. See the ASPRS Web site (http://www.asprs.org/) for the latest update on satellite information.

Let us discuss the various aspects of remotely sensed images the end user should be familiar with in selecting the right data source. Imagery can be expressed in four dimensions: (1) spatial, (2) spectral, (3) radiometric, and (4) temporal. In the following sections, we will discuss each of these aspects and the importance of each of the dimensions on information extraction from a scene.

2.1 SPATIAL RESOLUTION

Spatial resolution is often expressed in terms of ground sampling distance (GSD) and refers to the area covered on the ground by a single pixel. Spatial resolution is based on various factors, such as the sensor field of view (FOV), altitude at which the sensor is flown, and the number of detectors in the sensor, etc. Further, the spatial resolution of the sensors varies with the off-nadir viewing angle and is also influenced by the terrain on the ground.

Aerial sensors can collect imagery at varying resolutions by flying at different altitudes. There are two types of aerial cameras that exist in the market today: (1) fixed frame and (2) linear array sensors. For the fixed-frame sensors, the rectangular array of CCD detectors defines the swath width as well as height of the scene imaged

TABLE 2.1
Aerial Sensor Comparison

Sensor	Company	Sensor Type
DMC	Z/I	Frame
UltraCam	VeXcel	Frame
ADS40	Leica Geo Systems	Linear array
DSS	Applanix	Frame
DIMAC	DIMAC Systems	Frame

based on the aircraft altitude above the ground. Table 2.1 lists some of the popular digital aerial sensors in the industry today and the type of sensor.

On the other hand, satellites have a fixed orbit and a fixed spatial resolution at nadir. The final spatial resolution is influenced by the terrain differences as well as off-nadir viewing angles. Table 2.2 summarizes the orbital height of some of the sensors and associated spatial resolutions.

Although different terms are used in the industry to refer to types of spatial resolution, the following are some of the rough guidelines for definitions of spatial resolution: (1) low resolution is defined as pixels with GSD of 30 m or greater resolution, (2) medium resolution is GSD in the range of 2.0–30 m, (3) high resolution is GSD 0.5–2.0 m, and (4) very high resolution is pixel sizes < 0.5 m GSD.

Spatial resolution plays an important role in the objects that can be identified using remote sensing imagery. The National Imagery Interpretability Rating Scale (NIIRS), which originated in the U.S. intelligence community, is the standard used by imagery analysts and can be used as a reference on features that can be identified at various spatial resolutions. The NIIRS classification scheme has 10 levels, and Table 2.3 lists some of the features that can be identified at various levels.

TABLE 2.2
Satellite Sensor Comparison

Satellite	Orbital Height (km)	Spatial Resolution
Landsat	700	30 m MS, 15 m pan
SPOT 2, 4	832	20 m MS, 10 m pan
SPOT 5	832	10 m MS, 20 m SWIR, 2.5 m pan
IRS P5	618	2.5 m pan
CBERS	778	80 m MS, 160 m thermal
IKONOS	681	MS 4 m, pan 1 m
OrbView3	470	MS 4 m or pan 1 m
QuickBird2	450	MS 2.44 m, 0.61 m pan

TABLE 2.3
NIIRS Levels

NIIRS Level	Description
Level 0	Interpretability of the imagery is precluded by obscuration, degradation, or very poor resolution
Level 1	Detect a medium-sized port facility; distinguish between runways and taxiways at a large airfield
Level 2	Detect large buildings (e.g., hospitals, factories); identify road patterns, such as clover leafs, on major highway systems; detect the wake from a large (e.g., greater than 300 ft) ship
Level 3	Detect individual houses in residential neighborhoods; identify inland waterways navigable by barges
Level 4	Identify farm buildings as barns, silos, or residences; detect basketball courts, tennis courts, and volleyball courts in urban areas
Level 5	Detect open bay doors of vehicle storage buildings; identify tents (larger than two-person) at established recreational camping areas; distinguish between stands of coniferous and deciduous trees during leaf-off condition
Level 6	Identify automobiles as sedans or station wagons; identify individual telephone/electric poles in residential neighborhoods
Level 7	Detect individual steps on a stairway
Level 8	Identify grill detailing and/or the license plate on a passenger/truck type vehicle
Level 9	Detect individual spikes in railroad ties

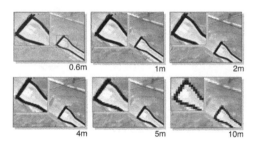

FIGURE 2.1 (Color figure follows p. 82.) Spatial resolution comparison.

Figure 2.1 shows a comparison of an airport runway at various spatial resolutions.

One of the key factors to keep in mind is that as the spatial resolution increases, the associated file size increases. Table 2.4 shows the uncompressed file sizes of 4-band 8-bit imagery at various resolutions for a 1-km² area.

Different applications require different spatial resolutions. For applications such as large area change detection, it is economical to use medium-resolution imagery with large swath widths, such as SPOT and Landsat, to cue into areas where changes of interest have occurred, and supplement high-resolution imagery to identify actual changes. Similarly, for planimetric applications, it is recommended that imagery

TABLE 2.4
Spatial Resolution
versus File Size

GSD	File Size (kb)
30 m	4
15 m	17
10 m	39
5 m	156
2.5 m	625
1 m	3906
0.6 m (2 ft)	10851
0.3 m (1 ft)	43403
0.15 m (6 in.)	173611

with the highest possible resolution be used to extract various features such as pavements, roads, etc.

Also, keep in mind that the spatial resolution and swath width go hand in hand. Typically, the higher the resolution, the smaller the footprint of the image. Figure 2.2 shows a comparison of swath widths of various satellites.

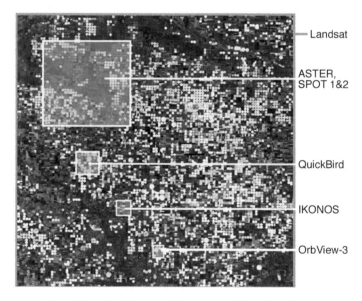

FIGURE 2.2 Swath widths of various satellite sensors. (From http://www.asprs.org/.)

2.2 SPECTRAL RESOLUTION

Spectral resolution refers to the number of spectral bands on a given sensor. Most of the aerial and satellite sensors image in the visible and infrared regions of the EM spectrum. Figure 2.3 shows a comparison of spectral bands of various sensors.

Depending on the number of bands in the sensors, various terms such as multi-, super-, and hyperspectral are used in the remote sensing industry. Commonly used definitions in the industry state that multispectral sensors have less than ten bands, superspectral sensors have bands greater than ten bands, and hyperspectral sensors usually have bands in hundreds.

For geographic information system (GIS) applications, where the image is used as a backdrop, 3-band natural color imagery in RGB format is sufficient. For information extraction products such as impervious surfaces, vegetation classification, etc., visible and near-infrared (VNIR) bands might be sufficient. For applications such as mineral exploration, forest species classification, etc., superspectral or hyperspectral data are required. In later chapters, we will also discuss data reduction techniques, such as principal component analysis (PCA), to handle large hyperspectral as well as superspectral data sets.

Understanding the spectral behavior of features that need to be extracted is essential in picking the right data source. There are several public domain sources for spectral signatures of various features, such as the Laboratory for Applications of Remote Sensing (LARS) at Purdue University, the University of California–Santa Barbara, and others. These institutions have compiled an extensive collection of a variety of features ranging from road materials and rooftops to vegetation signatures.

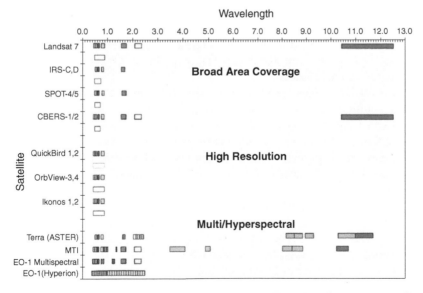

FIGURE 2.3 Spectral bands of various satellite sensors. (From http://www.asprs.org/.)

2.3 RADIOMETRIC RESOLUTION

The reflected signal is captured as an analog signal and then converted to a digital number (DN) value or a gray-level value. This process is referred to as analog-to-digital (A/D) conversion. *Radiometric resolution* is defined as the number of gray levels that can be recorded for a given pixel. A radiometric resolution of 8 bit will result in pixel values ranging from 0–255 and an 11-bit radiometric resolution can capture DN values ranging from 0–2047. The value range can be computed using the equation

$$N = 2^R, \tag{2.1}$$

where N is the range and R is the radiometric depth.

The radiometric range of the data captured is typically smaller than the designed sensor range. This is primarily because of sensor settings to ensure that data are not saturated at lower as well as higher ends. The lower end of the radiometric range is defined by the dark current of the CCD detectors, whereas the higher end is defined by A/D saturation. You have to keep in mind that 11-bit sensor can saturate at 1900–2000 DN range instead of 2048. The range of data captured in satellites is typically controlled by time delay integration (TDI), and for aerial sensors by fStop, which is the ratio of aperture to focal length. It is important to know the saturation levels of the sensors because information is lost in the saturated pixels.

Standard imagery formats store data in increments of 8 bits. Data captured at 11-bit or 12-bit resolution is stored in a 16-bit format with empty bits packed at the end. When displaying these files, if no stretch is applied, some of the common GIS software assume the data range to be 16 bit and display the images as black. The general rule of thumb is that the higher the radiometric resolution, the more information content is captured in the scene. Figure 2.4 shows 11-bit data enhanced in the shadow region, showing the details captured. Table 2.5 shows the radiometric resolution of various sensors.

The pixel DN values are affected by spectral response of the object, ground spatial resolution, feature size, neighboring features, and other factors. Further, you have to keep in mind that different spectral bands are captured at varying exposure times, which requires radiometric correction of the data for image analysis. This is very important when developing algorithms that need to be consistent over scenes collected under differing atmospheric conditions and over different geographic areas.

Although it is ideal to work in the reflectance space to develop consistent algorithms, there are a few software such as MODTRAN, ATCOR, and FLAASH that can convert DNs into reflectance space. Most of the sensors provide information on converting DNs into top-of-the-atmospheric radiance of various bands. This step compensates for the various integration and exposure times of different bands and converts the DNs in various bands to a consistent base. I recommend that the end user ask the data providers to provide these conversion factors. For example, DigitalGlobe has published the conversion factors as well as equations to convert DNs into radiance on its Web site.

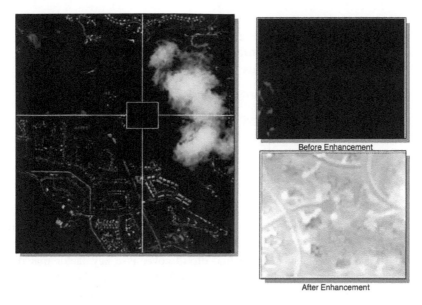

Before Enhancement

After Enhancement

FIGURE 2.4 (Color figure follows p. 82.) Importance of radiometric resolution.

TABLE 2.5
Sensor versus
Radiometric Resolution

Sensor	Radiometric Resolution (bits)
Landsat	8
IRS	7
SPOT	8
QuickBird	11
IKONOS	11
OrbView3	11
ADS40	12
DMC	12
UltraCam	12
DSS	12

2.4 TEMPORAL RESOLUTION

Temporal resolution refers to the time frequency when the same area of interest (AOI) is covered by the sensors. Imaging satellites are typically launched in a sun-synchronous orbit that results in the satellite revisiting a given spot on the earth at the same solar time. Further, sensors, such as IKONOS, QuickBird, OrbView3, and others have the flexibility to shoot off-nadir, increasing the frequency of the revisit.

TABLE 2.6
Temporal Resolution of Various Satellites

Satellite/Sensor	Revisit (days)
Landsat	8
IRS	7
SPOT	8
QuickBird	11
IKONOS	11
OrbView3	11

Also, at the higher latitudes, the frequency of revisits increases as compared to the equator. Table 2.6 shows the temporal resolution of various satellite sensors.

It is important to keep in mind that cloud cover is a key factor that dictates whether you can get a new cloud-free scene over a given area. While satellites are limited by the orbit, aerial sensors have the advantage of collecting data underneath the clouds. Constellation of satellites such as RapidEye, WorldView, and others will result in almost daily coverage of any place on the globe.

2.5 MULTISPECTRAL IMAGE ANALYSIS

Multispectral image analysis techniques are designed to exploit the spectral responses of various features in different spectral bands. Different features tend to have different spectral responses in various bands, as shown in Figure 2.5.

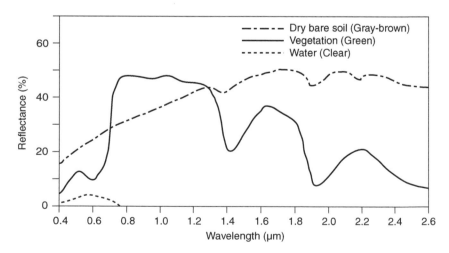

FIGURE 2.5 Spectral signatures. (From http://geog.hkbu.edu.hk/virtuallabs/rs/env_backgr_refl.htm.)

Bare soil tends to have a constant slope between various bands as compared to vegetation that has a steep slope between red and near-infrared (NIR) bands. Indices such as the Normalized Difference Vegetation Index (NDVI) exploit the slope of the spectral signatures of vegetation. Also, NIR reflectance value of healthy vegetation in the NIR band is 50% or greater, whereas vegetation reflectance in red band is less than 10%. Deep water bodies are characterized by low spectral responses in the VNIR range. Further knowledge of the phenomenology of the feature being classified, such as spectral responses of crops at various growth stages, is helpful in image analysis. Tassel cap coefficients are an example of a technique designed to reflect the capture of the phenomenology of changing vegetation in a growing season.

Traditional image analysis techniques are typically pixel-based techniques and classify each pixel in the image without regard to neighboring pixels. Metrics such as texture measures try to incorporate some of the contextual logic in image analysis but are restricted to a rectangular window. Some of the techniques also take advantage of the ancillary data to enhance feature extraction. In the coming chapters, we will discuss various techniques of image analysis and apply them to objects.

3 Why an Object-Oriented Approach?

Before we answer the question "Why object-oriented technology?" we will discuss the differences between pixel-based versus object-based image analyses using a practical application of multispectral remote sensing imagery. Let us take a look at the example of feature extraction of airport runways from imagery to demonstrate the point. Runways usually contain a variety of spectral signatures comprised of bright white paint used for runway markers and numbers, concrete runways, and black tread marks of planes landing. Traditional pixel-based supervised classification approaches require the creation of training sets to include the different spectral signatures of various features within the runway. Considering the advantages of an object-oriented approach to runway extraction, imagine if we somehow managed to create objects that defined the runway and features contained within the runway, then the problem would be simplified to classifying a few polygons rather than hundreds of pixels that are within the runway. Also, we can take advantage of the shape features, such as minimum size of the objects, to eliminate island pixels, that is, pixels that are entirely different as compared to the surrounding pixels. Also, we can use the same supervised classification technique used on pixels to classify the objects. Figure 3.1 demonstrates the results of feature extraction using the object-oriented approach.

The objects domain offers more dimensions for image analysis and can take advantage of some of the geographic information system (GIS) analysis. In the following sections, we will explore various facets of object-oriented approach for image classification.

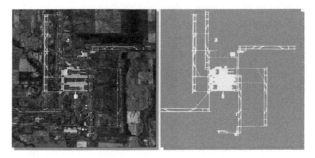

FIGURE 3.1 (Color figure follows p. 82.) Supervised runway classification based on objects. In yellow are the extracted airport features.

TABLE 3.1
Object Properties

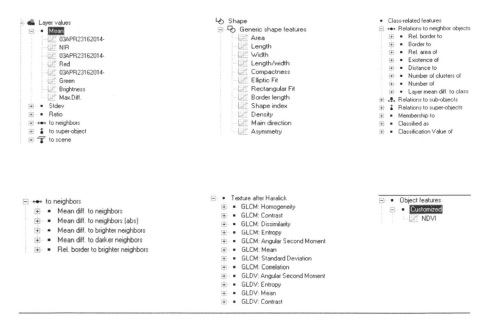

3.1 OBJECT PROPERTIES

Objects can be associated with hundreds of properties, such as shape, size, spectral response and others that can be used for image analysis. Table 3.1 lists some of the parameters available in eCognition software that can be used for classifying objects.

Most of these parameters are specific to object-based approaches and cannot be used in pixel-based image analysis. For example, in the object-oriented paradigm, you can use the size and shape of water bodies to categorize them into streams, lakes, and reservoirs; and computing similar size and shape parameters in pixel-based approaches requires additional techniques, such as region growing and edge enhancement. Figure 3.2 shows the results of comparing both paradigms for the runway feature extraction example discussed in the earlier paragraphs.

Although the results from the object-oriented approach look much cleaner, similar results can be obtained for the pixel-based approach by using techniques such as cluster busting, masking out large bare soil patches, and other GIS analysis to achieve results similar to the object-oriented approach. Figure 3.3 shows the spectral confusion of various impervious surfaces, such as roads and rooftops, with varying materials and changing spectral responses due to aging features.

Extracting these impervious surfaces purely based on spectral signatures is a daunting task, as there is considerable overlap between the spectral signatures. This is because of different roofing materials on homes ranging from shingles to brick

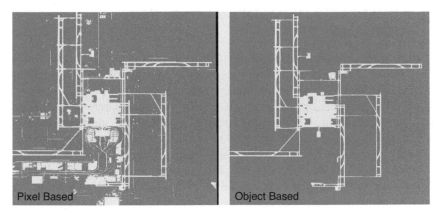

FIGURE 3.2 (Color figure follows p. 82.) Comparison of pixel-based versus object-based analysis.

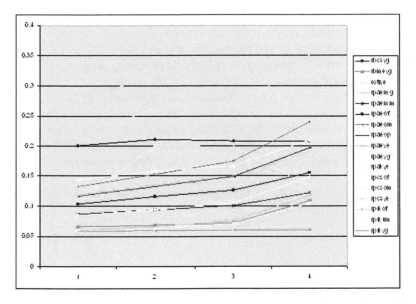

FIGURE 3.3 (Color figure follows p. 82.) Spectral signatures of various roof materials and aged road pavement materials. (From UCSB.)

to composite, and with varying color combinations. The problem is further complicated in areas where adobe rooftops are common and have similar spectral properties as bare soil. Additional information, such as shape, size, shadows, and other parameters, is required for extracting these impervious features. In later chapters we will use practical examples to demonstrate how to use various object properties to extract impervious surfaces.

3.2 ADVANTAGES OF AN OBJECT-ORIENTED APPROACH

Whereas spectral signatures of various features are typically expressed as single lines, in reality, features are made of a spectral family of members. In the object-oriented paradigm, you can take advantage of various spectral properties of objects, such as mean and standard deviation, to capture the spectral variations of features. You can further bring in derived information layers such as principal component analysis (PCA, edge-enhanced images, etc.) to add to the spectral dimensions.

Spatial properties, such as object area, length, width, and direction, can be exploited along with spectral properties. For example, we can differentiate between shadows and water using the knowledge that shadows are typically much smaller in area compared to water bodies.

Next, we look at the advantages of the object-oriented approach:

- Takes advantage of all dimensions of remote sensing, including the following:
  - Spectral (multispectral bands including panchromatic band)
  - Spatial (area, length, width, direction)
  - Morphological (shape parameters, texture)
  - Contextual (relationship to neighbors, proximity analysis)
  - Temporal (time series)
- Incorporates current proven methods/techniques used for image analysis such as supervised classification, fuzzy logic, and rule-based classification
- Incorporates some of the GIS functionality for thematic classification including the use of ancillary information, distance measures, etc.
- Extracts features from the same scene at different resolutions

Now, we are ready to begin the journey of exploring the new object-oriented paradigm for image analysis.

4 Creating Objects

Objects are the primitives that form an image. Image segmentation is one of the popular techniques used for creating objects from an image. Let us briefly discuss various software available for segmentation in the public domain. Using step-by-step examples, we will explore the creation of objects in eCognition.

4.1 IMAGE SEGMENTATION TECHNIQUES

Segmentation is the underlying concept for creating objects from pixels. The segmentation process involves dividing the image into regions or objects that have common properties. Typical image segmentation techniques involve one of the two processes: (1) region merging according to some measure of homogeneity and (2) separation of objects by finding edges using gradients of digital numbers (DNs) between neighboring pixels.

Region-merging approaches can be divided into two approaches: (1) region growing and (2) region split and merging. Region-growing technique involves pixel aggregation and starts with a set of seed points. From these seed pixels, the regions are grown by merging neighboring pixels that have similar properties. On the other hand, the region-splitting and -merging approach includes subdividing an image initially into a set of arbitrary, disjointed regions and then merging and/or splitting the regions based on the similarity rules for object creation.

The thresholding technique is a region-merging technique useful for discriminating objects from the background. In a simplistic case, images tend to have bimodal histograms and can be segmented into two different regions by simple thresholding pixels of gray-level values greater than a predefined threshold belonging to the scene and pixels with gray-level values smaller than the threshold as background pixels. Most thresholding techniques indirectly utilize the shape information contained within the image histogram. Image gray levels are divided into several subranges to perform thresholding.

Edge-based methods create objects based on contours of gray levels within an image. Watershed analysis is a popular edge-based image segmentation technique commonly used in the industry and uses the image gradient as input to subdivide the image into low-gradient catchment basins surrounded by high-gradient watershed lines. The watershed concept is based on the fact that a digital image can be interpreted as a topographic surface where the response of pixels or DNs can be assumed to represent elevation at that point. When this topographic surface is flooded from its minima and the waters coming from different sources are prevented from merging, the image is partitioned into two different sets: the catchment basins and the watershed lines. If this transformation is applied to the image gradient, the catchment basins correspond to the homogeneous regions of the image. The resulting catchment basins will consist of locally homogeneous connected sets of pixels. The watershed lines are made up of connected pixels exhibiting local maxima in gradient magnitude; to achieve a final segmentation, these pixels are typically absorbed into adjacent catchment basins.

Another edge-based technique is the connectivity-preserving relaxation-based segmentation method, called the *active contour model*, that starts with an initial boundary shape represented in the form of spline curves and iteratively modifies the boundary by applying various internal energy constraints such as shrink or expansion operations. The active contour evolves under the influence of both gradient and region information. The region information used is computed as the difference between the DN distribution in regions inside and outside the active contour. Another type of active contour model employs a Bayesian approach in which the likelihood function combines both the region and edge information. If you are interested in learning more about the image segmentation, these techniques are commonly used in the field of computer vision, medical imaging, video streaming, and other fields.

4.1.1 PUBLIC DOMAIN IMAGE SEGMENTATION SOFTWARE

There are several public domain software available for image segmentation. Most of the software is distributed by academic institutions. The following institutions have segmentation software available on their Web sites:

1. Rutgers University[1] has a nonparametric approach that uses mean shift algorithm for estimating density gradients for clustering.
2. The Penn State Engineering[2] technique uses a construct of pixel–pixel pairwise similarity matrix W based on intervening contours. The image is segmented with normalized cuts.
3. The University of Florida range image segmentation comparison project[3] uses a region segmentation algorithm for creating objects.
4. The UCSB Web site[4] discusses a technique that uses variational image segmentation, which uses image diffusion based on edge vector that uses color and textural properties of the image.

NASA's Goddard Space Flight Center has developed a technique for image segmentation called recursive hierarchical image segmentation,[5] and the source code is available for limited distribution. A Markov random chain–based image segmentation code is available from the Institute of Informatics in Hungary.[6] Bayesian-based image segmentation software codes can also be found at Purdue University.[7]

4.1.2 ECOGNITION SEGMENTATION

In this book, we will primarily use eCognition, a commercial software developed by Definiens (www.definiens.com) of Munich, Germany. eCognition uses heuristic algorithm for image segmentation. The patented eCognition segmentation algorithm creates image segments based on the following four criteria:

- Scale
- Color
- Smoothness
- Compactness

These criteria can be combined in numerous ways to obtain varying output results, thus enabling the user to create homogeneous image objects at any chosen resolution. In the following section, using hands-on examples, we will learn the effects of these four parameters on creating objects.

The scale parameter determines the maximum allowed heterogeneity within an object. For a given scale parameter, heterogeneous regions in an image will result in a fewer number of objects as compared to homogeneous regions. The size of image objects can be varied by varying the scale parameter value. Although there is no direct correlation between the scale parameter and the number of pixels per object, the heterogeneity at a given scale parameter is linearly dependent on the object size. Object homogeneity is defined by the homogeneity criterion and is represented by three criteria: color, smoothness, and compactness. Color is the most important criteria for creating meaningful objects. Color, smoothness, and compactness are variables that optimize the object's spectral homogeneity and spatial complexity. Understanding the effects of each of these criteria is required to segment the image and create objects for a given application. The color parameter defines the overall contribution of spectral values to define homogeneity. The shape factor is defined by two parameters: smoothness and compactness. The smoothness factor can be used to optimize image objects for smoother borders, and compactness can be used to optimize image for compactness. Please see *eCognition User Guide 4* to understand the science behind all four criteria. Now let us create objects using eCognition software.

EXAMPLE 4.1: CREATING OBJECTS AT DIFFERENT SCALES

Step 1: Import example1.tif.

Import the image into eCognition by using the menu option: Project → New or by using the import icon.

Make sure that the Files of type is set to All Raster Files to ensure all types of imagery files are displayed in the window as shown in Figure 4.1.

FIGURE 4.1 eCogniton raster import window.

FIGURE 4.2 Import window.

eCognition can accept a variety of common imagery formats accepted in the remote sensing industry such as GeoTIFF, PCI Pix format, ERDAS .img format, ER Mapper .ECW format, and others. Please see *eCognition User Guide 4* for the detailed list of image formats that can be imported into eCognition.

Browse to the folder with example imagery and select example1.tif and the screen shown in Figure 4.2 will appear.

eCognition automatically reads the header information from the GeoTIFF header, including the projection and pixel GSD, and imports all the spectral bands in a given file. It is important to make sure the units are assigned to meters. Although eCognition allows you to work in pixel units, changing the units to meters will help you export the results, raster as well as vector, with geographic coordinates, and the output from eCognition can be readily integrated into other GIS applications. Even if you worked in pixel units, you always have the option of using software such as ArcCatalog from ESRI to attach geographic information to the output layers from eCognition.

The example1.tif is a QuickBird 2 (www.digitalglobe.com) image and has four multispectral bands that represent the visible and near-infrared (VNIR) bands. When importing new image layers in eCognition, information on the radiometric resolution of various layers are displayed along with the geographic coordinates. At this stage, you have the option of adding additional image layers or removing some of the layers for image analysis. QuickBird imagery is distributed with bands arranged in the following sequence: blue, green, red and near-infrared (NIR), respectively. You have the option of creating aliases for each of the layers at this stage or using the menu option within eCognition. It is important to note that it is possible to bring in other layers with differing spatial and radiometric resolutions into eCognition simultaneously.

Now click on the Create button.

The image will appear bluish in color as shown in Figure 4.3.

FIGURE 4.3 Default display in eCognition.

This is because of the default band combination picked by eCognition for display. For any image with more than two bands, eCognition automatically assigns band 1 to red, band 2 to green, and band 3 to blue color display palettes, respectively. Currently, the remote sensing industry is trying to define a standard format for distributing imagery, as different commercial data providers follow different standards for imagery distribution. Let us change the band combinations in this exercise to display the natural color image.

Step 2: Set the band combinations using the icon Edit Image Layer Mixing or by navigating to View → Edit Image Layer Mixing option in the menu and reassign the band combinations as shown in Figure 4.4.

FIGURE 4.4 Image layer mixing menu in eCognition.

The default stretch applied for display in eCognition is the Linear Stretch option. You can enhance the image display using various options including the Standard Deviation Stretch, Gamma, or Histogram Stretch. Try these different enhancing algorithms to see the effects on visual display of the image. Displaying example1.tif(4) as red, example1.tif(3) as green, and example1.tif(2) as blue will result in a false color composite (FCC).

Preset options allow you to display the image as gray scale, in a 3-band combination, and as a 6-layer mix. The Shift keys assign different spectral band combinations to RGB. If you set the preset mode to the 6-layer mix, the Shift keys assign different weights to bands. These are effective visualization tools that should be used to identify various features before starting a project, especially for change detection studies.

Step 3: Segment the image at a scale parameter of 10.

Click on the icon Multiresolution Segmentation or navigate to menu option Image Objects → Multiresolution Segmentation.

Use the default parameters for segmentation as shown in Figure 4.5.

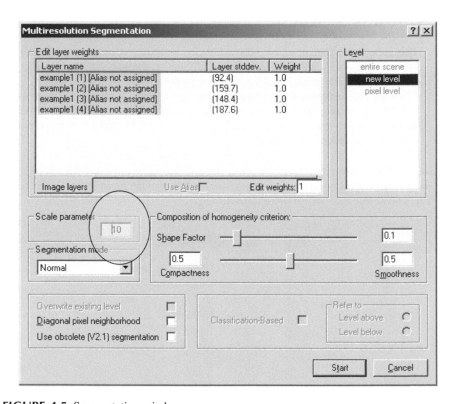

FIGURE 4.5 Segmentation window.

FIGURE 4.6 (Color figure follows p. 82.) Object boundaries.

Click on Start to begin segmentation.

To see the object boundaries, click on the icon Show/Hide Outlines. This will display the object boundaries of the segmented image as shown in Figure 4.6.

Zoom to various regions in the scene to observe how the segmentation process created the object boundaries. You will notice that homogenous areas have large objects, whereas heterogeneous areas have smaller objects. Click on any object and it is highlighted in red color. eCognition provides a unique environment where you are working in vector as well as raster domains.

Step 4: Segment the image at a scale parameter of 50 on the Level 1 and notice the size of the objects and the corresponding object boundaries.

Step 5: Repeat step 4 and segment the image at scale parameters of 75 and 100.

You will notice that the increasing scale parameter will result in a fewer number of objects. Let us compare the object boundaries of heterogeneous areas versus homogenous areas at different scale levels. Figure 4.7 shows a comparison of object boundaries of a homogeneous bare soil patch with a heterogeneous wooded area next to it.

As expected, the bare soil patch results in fewer objects with decreasing scale parameters as compared to heterogeneous wooded areas next to it.

In this example we have learned to:

- Segment images in eCognition and create objects
- Display images using various image enhancement techniques
- Visualize the object boundaries
- Create objects at various scale parameters
- Observe the effects of scale parameter on homogeneous versus heterogeneous objects

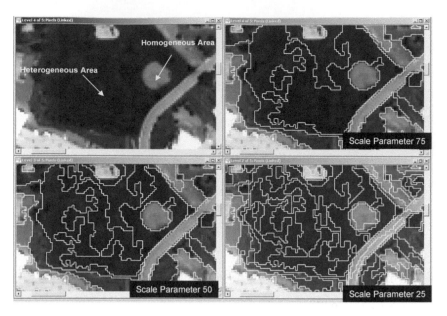

FIGURE 4.7 Scale parameter impact on homogeneous versus heterogeneous areas.

Use the icon Pixel or Transparent View to see segmented raster displayed as object versus actual raster. If you want to change the outline color of the object boundaries, click on the icon Edit Highlight Colors and select a different color as shown in Figure 4.8.

Try different scale parameters to see their effects on object size and corresponding features that can be extracted. Figure 4.6 shows an example of an image segmented at different scales. eCognition processing time is dependent on the number of objects created. The number of objects is displayed on the bottom right corner of the eCognition window. Observe the relationship between the number of objects created based on the image scale. Also, experiment with segmenting images at larger scale parameters until the whole scene becomes one object.

Table 4.1 shows an example of scale parameter versus number of objects for a test scene of 1200×1200 pixels with 8-bit resolution and corresponding eCognition file size.

Notice the significant drop in the number of objects as the scale parameter increases and the corresponding file size decreases.

eCognition creates numerous temporary files during the segmentation process. For large images, it is recommended that a disk drive with 100 GB or more be used for temporary file space. eCognition gives the user the option of changing the path to where temporary files are stored. It can be accomplished by navigating to Help System Info → Temp Path and entering the new directory name.

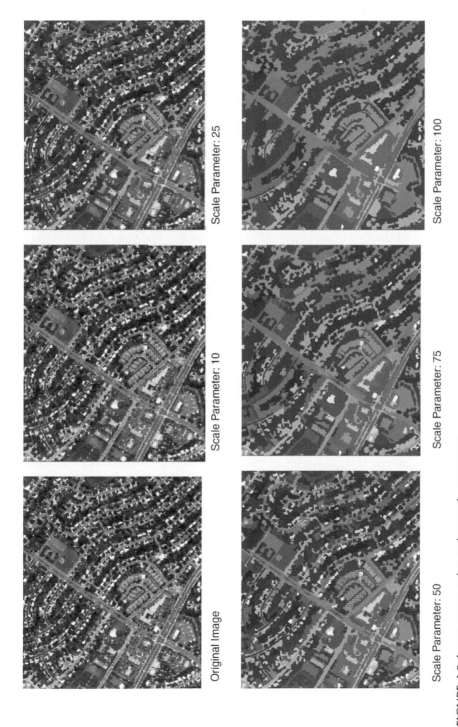

Scale Parameter: 25

Scale Parameter: 100

Scale Parameter: 10

Scale Parameter: 75

Original Image

Scale Parameter: 50

FIGURE 4.8 Image segmented at various scale parameters.

TABLE 4.1
Scale Parameter versus Number of Objects

Scale Parameter	Number of Objects	File Size (MB)
10	2E+05	36
25	27082	8
50	6911	3
75	3045	2
100	1658	1.6
500	50	0.5

4.2 CREATING AND CLASSIFYING OBJECTS AT MULTIPLE SCALES

One of the advantages of the object-oriented approach is that you can create objects at various scales and extract features at different resolutions. A practical example of such an approach would be classifying a large water body, such as the ocean, at scale parameter of 250 and classifying other objects on land at much smaller scales. eCognition can be used to construct a hierarchical network of image objects that represents the image information at different spatial resolutions simultaneously as shown in Figure 4.9. The image objects are networked so that each image object can be related to its neighborhood objects on a given level, to its parent objects at higher levels, as well as to its subobjects in levels below. The objects A and B on the topmost level, Level 4 in the illustration, are the parent objects. The area represented by a parent object is the sum of the areas of subobjects. Area A = Area A1 + Area A2, Area A1 = Area 1.1 + Area 1.2, and so on. Object A1 and Object A2 boundaries will be within the object boundaries of parent object A. This is a powerful concept that we will explore in later chapters. This logic also works the other way, bottom layer to top layer, where top-level objects are constructed based on the subobjects in the layer below. Subobjects are merged together to form larger parent objects in the next higher level.

To preserve this network and exploit the object relationships, it is recommended that decreasing scale parameters be used for segmentation at lower levels. eCognition will not allow for levels below a given level to be created with increasing scale parameters. Also, eCognition allows you to have multiple levels segmented with the same scale parameters for applications such as change detection. Let us do a few examples that will get us acquainted with the basic functionality and features of the eCognition software.

EXAMPLE 4.2: CREATING A PROJECT IN ECOGNITION

Step 1: Create a new project.

Step 2: Import example2.tif.

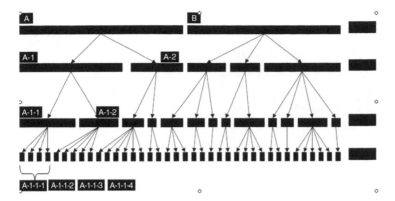

FIGURE 4.9 Parent–child relationships within eCognition.

Step 3: Make sure the units are set to meters.

Step 4: Set the temporary path to a directory with plenty of disk space.

Step 5: Click on the Edit Layer Mixing icon to select band combinations.

Step 6: Segment the image at a scale parameter of 100.

Step 7: Click on the Show/Hide Outlines icon to see the object boundaries created.

Step 8: Click on Image Object Information 1 icon and select any object to see the associated properties. Use the Pixel or Transparent View icon to toggle between objects and imagery mode.

Step 9: Click on the Feature View icon to display the Features window.

Step 10: Click on the Class Hierarchy icon to display the Class Hierarchy window.

Step 11: Select Window → New Window to open a second window.

eCognition lets us open multiple windows and link them.

To organize the layout, Use Windows → Tile Horizontally and select Windows → Link All Windows to geographically link the windows.

It is recommended that a typical layout as shown in Figure 4.10 be used for exercises in this book. This layout will allow you to display the thematic map in one window and the actual image in the second window. Also, while working on multiple levels, you can simultaneously display multiple levels in different display windows and have them linked.

Step 12: Use Toolbars + Default Settings → Save GUI Settings as Default. This will enable the same layout for future eCognition projects.

FIGURE 4.10 Suggested layout in eCognition.

Step 13: Save the project.

You will notice that eCognition projects have two files, a .dpr file, which is the project file, and an associated .dps file that saves the project settings.

In this example we have learned to:

- Segment images and create objects in geographic projection
- Display images in the right combination
- Visualize object boundaries in vector space
- Look at various object properties that can be used for image analysis

4.3   OBJECT CLASSIFICATION

To demonstrate the use of traditional image analysis techniques on objects, let us create a thematic map using the nearest-neighbor supervised classification technique. Using the functions within eCognition, we will select training data sets for various classes. For this exercise, let us use a subset of Landsat scene and perform a simple image classification.

EXAMPLE 4.3: FIRST OBJECT-BASED IMAGE CLASSIFICATION IN ECOGNITION

Step 1: Create a new project.

Step 2: Import example3.tif.

FIGURE 4.11 Assigning no data values.

Step 3: Ensure the units are in meters.

Step 4: Let us set a null value by navigating to Project → Assign No Data Value.

You have the option of setting null values for individual bands or one value for all bands. Let us set 0 as a null value for all the bands as shown in Figure 4.11.

Step 5: Segment the image at a scale parameter of 10, using the default parameters for shape.

Step 6: Now let us create the classes for image analysis.

eCognition stores the classes in a class hierarchy file that can be saved as part of the project or can be exported as a separate file. Class hierarchy files have a file extension .dkb. Let us import a class hierarchy that I have already created. Navigate to Classification-Load Class Hierarchy. Browse to the example 3 folder and select the file "example3.dkb." We will learn in the following sections how to use "alias" functions so that class hierarchy files can be used in other projects. Make sure the Class Hierarchy window is open. The following classes should appear in the Class Hierarchy window: Bare Soil, Built-up, Vegetation, and Water. Currently, these classes are empty and do not contain any rules for image classification. We will discuss in detail the tabs on the bottom of the Class Hierarchy window in the coming chapters. Also, you can create a new class either by right-clicking in the Class Hierarchy window and selecting Insert Class or by navigating to Classification → Edit Classes → Insert Class. The class hierarchy should look as in Figure 4.12.

FIGURE 4.12 Class hierarchy file.

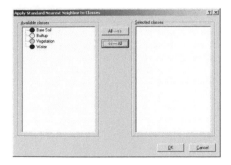

FIGURE 4.13 Applying standard nearest-neighbor classification rule.

Step 7: Let us add the nearest-neighbor classification rules for all of the classes.

This can be accomplished by navigating to Classification → Nearest Neighbor → Apply Standard NN to Classes. Make sure you add all classes into the right-hand window. If you are interested in only applying standard NN to only a few classes, you can do so only by adding a select few classes in the right-hand window.

For this example, click on All to make sure all the classes appear in the right-hand window. You can unselect classes by selecting the "←- All" button as shown in Figure 4.13.

Step 8: Now, let us examine the rules created for the nearest-neighbor classification.

Double-clicking on a given class will open up the Class Description window. Let us examine the Water class. Figure 4.14 shows the Class Description window with rules automatically added to the Water class.

The nearest-neighbor rule uses all four multispectral bands for image classification. You will notice the same rule has been applied to the other three classes in the class

FIGURE 4.14 Nearest-neighbor classification rule.

hierarchy. You could also add the nearest-neighbor rule by double-clicking on the and[min] option and selecting the Standard Nearest Neighbor option. If you are wondering about various features within the Class Description window, we will discuss, in detail, features such as Contained, Inherited, and and[min] in the subsequent chapters. The difference between Standard Nearest Neighbor and Nearest Neighbor techniques is that the variables used for standard nearest neighbor apply to all the classes within the project, whereas you can modify the variables based on features in the Nearest Neighbor option. If you want to change the variables selected for Standard Neighbor or the Nearest Neighbor options, right-click on the variable in the Class Description window and select the Edit Expression option. This will open the Edit Nearest Neighbor Feature Space window and will allow you to select other variables or delete existing variables.

Step 9: Let us create training samples for each of the four classes in the class hierarchy files.

To create training samples, use the three sample selection icons available in the menu. The first icon will allow you to select the samples from the main display window. The second icon will open a window that shows the feature values of the samples selected, and the third icon shows a table of values that describe the closeness of samples selected to a given class. To start picking the samples, click on icon 2 and icon 3 of the sample menu, and arrange the windows so that you can see the values.

You can also access these windows through menu selection by navigating to Samples.

Let us first pick the samples for the class Water. Select the Water class in the Class Hierarchy window and pan to the top left of the display window where a river is present. Double-clicking in the image window will allow you to select a sample. Please note that the sample space is in objects rather than in pixels. Figure 4.15 shows the samples I have picked for all the classes.

We have taken our first step into the object-oriented paradigm of image analysis. You can unselect samples by double-clicking on the object. The Sample Editor

FIGURE 4.15 (Color figure follows p. 82.) Training sample selection in object space.

window will display the spectral values of the selected samples picked. For every additional sample selected, the Sample Selection Information Window will update the information and shows how close a given sample is to all other classes in the class hierarchy. This can be used to select appropriate training samples for each class.

Repeat the above process by selecting samples for all other classes. It is recommended that samples be picked that are evenly distributed across the scene, and the number of samples is statistically significant to represent a given class. Also, displaying the image as FCC is helpful for picking samples for vegetation as it is displayed in a bright red color.

Step 10: Now that we have selected the training samples, we are ready to perform our first object-based image classification. Click on the Classify icon or navigate to menu option Classification → Classify to perform supervised classification.

The output should look similar to that in Figure 4.16.

Your output might look different and is based on the training samples you have selected. If you are not satisfied with the results, add more samples to each class and reclassify the image. Voila! You just created your first object-based thematic map. Use the Pixel-Transparency icon to switch between opaque and semitransparent display modes. In the semitransparent mode, you should be able to see the image features underneath the thematic map.

Step 11: Perform accuracy assessment.

I have dedicated a chapter to discuss various accuracy assessment procedures, but let us perform some basic accuracy assessment of the thematic map we just created. Let us use the eCognition built-in functionality to perform the accuracy assessment.

FIGURE 4.16 (Color figure follows p. 82.) First object-based thematic classification results.

To accomplish this, let us navigate to Tools → Accuracy Assessment. Select the option Error Matrix Based on Samples as shown earlier, and click on Show Statistics. The window shows the various accuracy metrics including overall accuracy and kappa coefficient. We have now completed the accuracy assessment. We will discuss various statistical measures in the accuracy assessment chapter.

Step 12: Save project.

In this example we have learned to:

- Import class hierarchy
- Apply nearest neighbor rules to classes
- Create object-based samples
- Perform supervised classification
- Perform accuracy assessment

We just employed an object-oriented paradigm to classify the image. Figure 4.17 shows a comparison of results of the same image using pixel-based supervised classification approach.

Notice that in the pixel-based image classification, on the left-hand side of the figure, pixels that are Bare Soil are classified as Built-up. Also notice that vegetation between the runway is classified as Bare Soil, whereas the object-based approach accurately identifies these features.

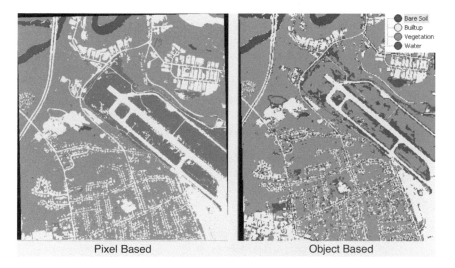

Pixel Based   Object Based

FIGURE 4.17 (Color figure follows p. 82.) Comparison of pixel-based and object-based supervised classification results.

4.4 CREATING MULTIPLE LEVELS

Let us explore the creation of objects at different scale parameters and at different levels in eCognition. This is a powerful functionality that will allow us to extract various features at different object sizes. A typical image is comprised of features of various sizes. Large homogeneous objects within the scene can be extracted at large object sizes, and heterogeneous features can be classified as smaller objects. In eCognition, the image can be segmented at different scale parameters. As mentioned in the previous section, it is recommended that start segmenting at large scales and create sublevels with segmentation with lower-scale parameter values. This will ensure the parent–child relationship between various objects at different levels. Also, eCognition has functionality that will allow us to explore relationships between different levels, parent and child object and vice versa. Let us look at some examples that explore this concept.

EXAMPLE 4.4: CREATING MULTIPLE LEVELS

Step 1: Create a new project.

Step 2: Import files example4-blue.tif, example4-green.tif, example4-red.tif, and example4-NIR.tif.

> These samples are from IKONOS-2 of space imaging (www.spaceimaging.com) at 1 m resolution. As in any Microsoft applications, you can use the Ctrl key for selecting multiple files and the Shift key to select a range of files.

Step 3: Segment the image using a scale parameter of 100.

Step 4: Segment the image at a scale parameter of 50, one level below as shown in Figure 4.18.

> You have the option of creating a new level below or above the current level. In eCognition, you have to use decreasing scale parameters for levels below and increasing values of scale parameters for levels above to preserve the parent–child

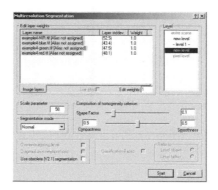

FIGURE 4.18 Creating a new level.

object relationships with the exception being that you can create levels above or below with the same scale parameter. The level with scale parameter 100 is now Level 2. Use the up and down arrow icons to toggle between the levels. Use the object boundary icon as well as the pixel-transparency icon to see the differences in object boundaries between the two levels.

Step 5: Segment the image at a scale parameter of 25, on the level below.

Using a similar approach, create a level below Level 1. This will now make the level with a scale parameter of 100 as Level 3 and a scale parameter of 50 as Level 2.

Step 6: Navigate to levels 2,3 using the up and down arrow icons.

Step 7: Go back to Level 1 and, using the Feature View window, navigate to Object Features → Layer Values → to Super Object → Mean Diff to Super Object.

Double-click on any of the bands to display the values of mean differences of child objects with parent object. Moving the cursor around in the display window will display the value of each object. Figure 4.19 below shows a comparison of a commercial neighborhood segmented at various scale parameters.

Notice that by selecting the parent object at Level 4, the boundary is preserved on all other sublevels. If our objective was to extract parking lots and buildings, we

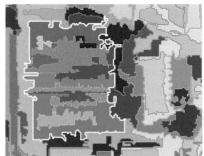
Level 3 (Scale Parameter:100)

Level 2 (Scale Parameter:50)

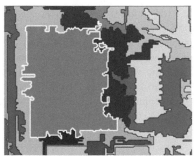
Level 1 (Scale Parameter:25)

Original Image

FIGURE 4.19 Multilevel image segmentation.

FIGURE 4.20 Comparing values of child object with super/parent object.

would extract them at Level 3 using the object outlined in yellow. If we need more information about the object, such as identifying driving lanes between parking lanes, we can use Level 2. At Level 1, we can identify groups of cars. Also, notice the building to the right of the parking lot and how subobjects are created at lower-scale parameters.

Step 9: To view relationships of objects between Level 2 and Level 3, navigate to Level 1 and use Feature View options to display values as shown in Figure 4.20.

Step 10: Repeat the steps by going to Level 3 and displaying object relationships between objects in Level 3 and Level 4.

eCognition allows you to explore the relationship between child objects and parent objects that are more than one level above. To demonstrate this functionality, navigate to Object Features → Layer Values → to Super Object → Mean Diff to Super Object. Select any band, right-click and select the option Copy Feature. A window will prompt you to enter the number of levels above the current level that you want to use to investigate the parent–child relationship. If you are on Level 1 and want to investigate the parent–child relationship with objects in Level 3, type in a value of 2; to investigate relationships between Level 1 and Level 4, enter 3, and so on. Let us say we created a feature that explores the relationship between Level 1 and Level 3; now the same feature can be used to investigate the relationship between objects in Level 2 and Level 4.

Step 11: Save project.

In this example we have learned to:

- Create multiple levels
- Explore object-to-object relationships between sub- to superobjects
- Create features that can investigate relationships in objects in multiple levels

We just took a giant leap into the object paradigm where we are now capable of creating objects and different levels. In later chapters, we will use this concept to extract various features at different levels and object sizes. As I mentioned before, object-to-object relationships are preserved between various levels, allowing us to extract features at different object sizes and still bring all of them together to create the final thematic map.

4.5   CREATING CLASS HIERARCHY AND CLASSIFYING OBJECTS

In Example 4.3, we imported a class hierarchy that was already created. In this section, we will learn how to create a new class hierarchy in eCognition and extract objects at different levels.

EXAMPLE 4.5: CREATING MULTIPLE-LEVEL CLASS HIERARCHY

Step 1: Import example5.tif.

This is a color-infrared (CIR) file that is from an ADS40 sensor from Leica Geo-Systems (www.leica.com).

Step 2: Set up the layout as discussed in Example 4.1.

Step 3: Segment the image at a scale parameter of 10.

Step 4: Segment the image at a scale parameter of 10 on a new level, one below current level.

Step 5: Right-click in Class Hierarchy window and select Insert New Class.

Step 6: Let us create a class hierarchy.

First, let us create a class called Level 1 so that any classes we create under Level 1 will apply to that level only. Double-click on and[min]. Using the Insert Expression window, navigate to Object Features → Hierarchy and double-click on Level. This will open up the Membership Function window and create a class rule as shown in Figure 4.21.

Select the Inverse T function by clicking on the third button in the second row under the Initialize option, and type the numbers 0 in the left border and 2 in the right border for the membership function. eCognition automatically computes the value

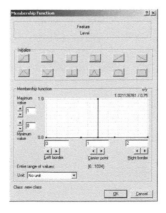

FIGURE 4.21   Creating a class in eCognition.

for center point, which in this case is 1. Let us leave the defaults for minimum and maximum values in the left-hand side of the membership function. As we are creating a rule for setting up rules that apply to a certain level, the default values of "no unit" will apply in this case. Also, notice that the field Entire Range of Values gives you an idea of the typical values you can use for defining the membership function. Click on OK to finish defining the membership function and then click on OK in the Class Description window to create the class Level 1.

We just created a hard rule that says the rule will apply only if the level is one. You could have recreated the same rule by using values such as 0.9 and 1.1 in the left and right borders that will result in the average value of 1 being assigned to the center point. The various functions/shapes under Initialize allow you to define sophisticated membership functions that we will explore in later chapters.

Similarly, let us create a new class called Level 2 and add rules that will ensure that the classes we create will apply to Level 2 only. Follow the same steps we used for creating the class Level 1 and, in the Membership Function window, type in values 0 and 4 for the left and right borders, respectively, which will result in the center point being assigned a value of 2. Try different combinations of left and right border values to ensure that the center point is assigned a value of 2. Also, note that we could have created Level 2 by copying Level 1. This can be accomplished by selecting Level 1 in the Class Hierarchy window and by using the right-click or Ctrl+C to copy the class as shown in Figure 4.22.

A new class named Copy of Level 1 will be created. Double-click on this and you will be able to change the name to Level 2 from Copy of Level 1. Double-click on and[min] to open the membership function and change the left and right border values to represent Level 2, as discussed in the previous paragraph.

Step 7: Let us now create two classes: Vegetation and Built-up.

In the Class Hierarchy window, right-click and select Insert New Class. Create a class called Vegetation. Change the color class to Green.

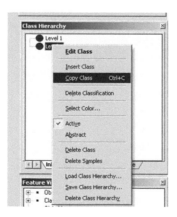

FIGURE 4.22 Copying class.

FIGURE 4.23 Multilevel class hierarchy.

Double-click on and[min] and select Standard Nearest Neighbor rule.

In Example 4.2 we performed this by navigating to Classification → Nearest Neighbor → Apply Standard NN to All Classes. Now we will manually set up this rule. Click on OK to complete creating the Vegetation class. Repeat the techniques in the previous two paragraphs to create Water class.

In this example, let us classify Vegetation on Level 2 and Water on Level 1. Using the mouse, drag Vegetation class under Level 2 in the Class Hierarchy window and drag Water class under Level 2.

This will ensure that any classes under the class Level 2 will apply for Level 2 only. The class hierarchy should look as shown in Figure 4.23.

Step 8: Go to Level 2 and select samples for Vegetation.

Step 9: Click on the Classify icon and only Vegetation will be classified.

Step 10: Go to Level 1 and select samples for Water.

Step 11: Perform the classification.

The results should look similar to those in Figure 4.24.

Step 12: Now, let us look at the class description of Water.

Because we have dragged Water under the class Level 1, you will notice that Water has inherited the rules from Level 1 class as shown in Figure 4.25.

So, even if you selected Vegetation samples on Level 1, the final classification will include results for Water only. This is a powerful functionality and can be used for two different purposes: (1) masking features we are interested in and (2) creating subclasses within the class. An example would be classifying all vegetation in a scene using normalized difference vegetation index (NDVI) and further creating subclasses, Trees and Lawns, under the Vegetation class. In the Class Description window, the rules under Contained will apply to that class along with any rules inherited from the parent class.

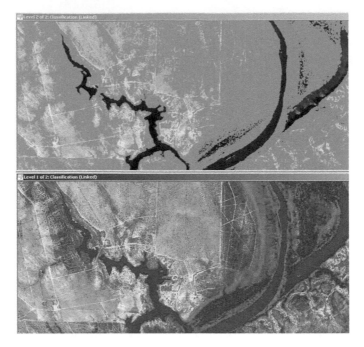

FIGURE 4.24 (Color figure follows p. 82.) Feature extraction on two different levels.

FIGURE 4.25 Inheritance of objects.

Step 13: Save project.

In this example we have learned to:

- Create classes
- Extract objects at different levels and at different sizes
- Understand the concept of inheritance in class hierarchy

You might be wondering how we can merge the results from Level 1 and Level 2 and create one layer with all the classes. The next section demonstrates how to get information from multiple layers and merge it into one.

4.6 FINAL CLASSIFICATION USING OBJECT RELATIONSHIPS BETWEEN LEVELS

Let us build on the multilevel classification we have done in the previous example to create a final classification on one level with both classes.

EXAMPLE **4.6:** PERFORMING MULTILEVEL FEATURE EXTRACTION AND CREATING ONE LAYER WITH FINAL CLASSIFICATION

Step 1: Open example4.5.dpr.

Step 2: Create a new level, one below Level 1.

Step 3: We have to modify the class description files of Water and Vegetation to reflect the new levels.

As we created a level below, Level 2 becomes Level 3, and Level 1 becomes Level 2. Let us modify the class hierarchy files to reflect that.

Step 4: Create 4 new classes — Level 1, Water Final, Vegetation Final, and Others Final. Navigate to Level 1.

Drag Water Final, Vegetation Final, and Others Final classes under Level 1. Let us add a rule that extracts Water objects from Level 2. Figure 4.26 shows how to add the rule.

Let us now add a rule to the Vegetation Final class. As discussed earlier, we are accessing objects from two levels above the current level. So, navigate to Class Related Features → Relation to Super Objects → Existence of Vegetation, right-click on the feature and select the option Copy Feature, and make sure we selected Feature Distance 2, because Vegetation was classified two levels above the level where we are creating the final classification as shown in Figure 4.27.

The last rule is to create Others Final, which should be everything that is not Water and Vegetation. Figure 4.28 shows the rule base.

Click on the Classify icon ensuring that the icon Classify with or without Features is set to the former option. We can use the default value of 1 for Number of Cycles option. We will discuss the annealing options in later chapters.

Step 5: Save project.

In this example we have learned to use parent–child relationships from multiple levels.

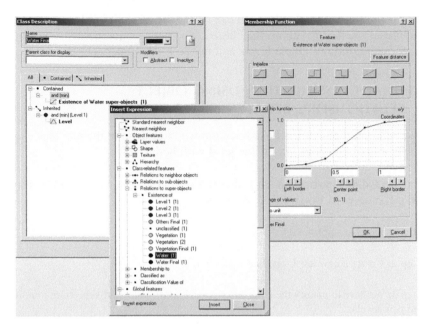

FIGURE 4.26 Setting up rule to extract thematic information from the previous level.

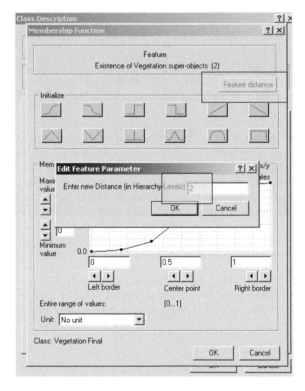

FIGURE 4.27 Accessing feature information from two levels above the current level.

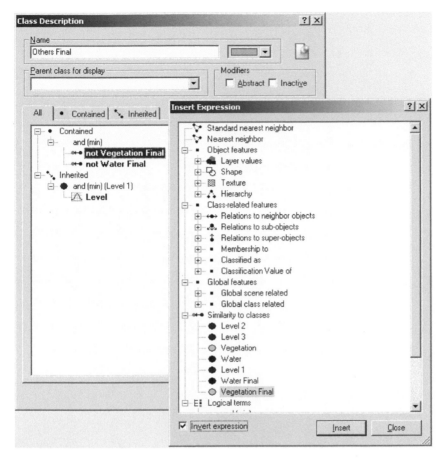

FIGURE 4.28 Using the inverse rule for classification.

Now, we are starting to explore the power of object-oriented classification. This approach can be used to identify large water bodies, mask out objects that are of no relevance, and explore this multilevel object paradigm for change detection. It is important to remember that if the objects on top levels are not merged, the objects below will not be merged even if contiguous objects have the same classification.

NOTES

1. http://www.caip.rutgers.edu/~comanici/segm_images.html
2. http://www.cis.upenn.edu/~jshi/software/
3. http://marathon.csee.usf.edu/range/seg-comp/SegComp.html
4. http://vision.ece.ucsb.edu/segmentation/
5. http://techtransfer.gsfc.nasa.gov/HSEG.htm
6. http://www.inf.u-szeged.hu/~kato/software/
7. http://dynamo.ecn.purdue.edu/~bouman/software/segmentation/

5 Object-Based Image Analysis

5.1 IMAGE ANALYSIS TECHNIQUES

In this section, we will briefly discuss various image analysis techniques commonly used in the remote sensing industry. I will cover various techniques, including unsupervised and supervised classification, rule-based classification approaches, neural net and fuzzy logic classification and, finally, classification and regression trees (CART) and decision trees. Knowledge of these techniques is useful in developing rule bases to extract features of interest from the image. Although these techniques have been employed for pixel-based analysis, I will demonstrate that these techniques can be easily extended to objects.

5.1.1 UNSUPERVISED CLASSIFICATION

Unsupervised classification techniques have been extensively used in image analysis for several decades. The objective of this technique is to group pixels with similar multispectral responses, in various spectral bands, into clusters or classes that are statistically separable. Cluster definition is dependent on the parameters chosen, such as spectral bands, derived spectral ratios, such as Normalized Difference Vegetation Index (NDVI), and other parameters. Each individual pixel within the scene or image is compared to each discrete cluster to see the closest fit. The final result is a thematic map of all pixels in the image, assigned to one of the clusters each pixel is most likely to belong. Metrics such as Euclidian, Bhattacharya distance, and others are used as a measure to find the closeness of a pixel to a given cluster. The thematic class or cluster then must be interpreted by the user as to what the clusters mean in terms of ground truth. This approach requires a priori knowledge of the scene and the content within the scene. The number of clusters can be modified based on the user's knowledge of features within the scene. One of the drawbacks of this technique is the generalization that can result in arbitrary clusters which do not have any correlation with features on the ground. Further, pixels belonging to clusters that have spectral overlap are often assigned to one of the classes based on a single metric with potential for gross misclassification errors.

There are several different types of unsupervised algorithms that are commonly used for image analysis. The two most frequently used algorithms are the K-means and the iterative self-organizing data analysis technique algorithm (ISODATA) clustering algorithms. Both of these algorithms are iterative procedures. The initial step for either technique includes assigning an arbitrary initial cluster vector of potential features within the scene. This is followed by assigning each pixel to the closest cluster. In the third step, the new cluster mean vectors are updated based on all the spectral values of the pixels within that cluster. The second and third steps are iterated until the difference between the successive iterations is small. The difference can be measured in several different ways, including measuring the distances of the mean cluster vector and other measures.

K-means algorithm minimizes the within-cluster variability. The objective function is the sums-of-squares distances (errors) between each pixel, and it is assigned to the cluster center.

$$SS_{distances} = \Sigma \, [x - C(x)]^2 \tag{5.1}$$

where $C(x)$ is the mean of the cluster that pixel x is assigned to.

Minimizing the $SS_{distances}$ is equivalent to minimizing the mean squared error (MSE). The MSE is a measure of the within-cluster variability.

$$MSE = \Sigma \, [x - C(x)]2/(N - c) \, b = SS_{distances}/(N - c) \, b \tag{5.2}$$

where N is the number of pixels, c indicates the number of clusters, and b is the number of spectral bands.

The ISODATA algorithm also includes functionality for splitting and merging clusters. Clusters are merged if either the number of pixels in a cluster is less than a certain threshold or if the centers of two clusters are closer than a certain threshold. Clusters are split into two different clusters if the cluster standard deviation exceeds a predefined value and the number of pixels is twice the threshold for the minimum number. The ISODATA algorithm is similar to the K-means algorithm with the primary difference being that the ISODATA algorithm allows for a different number of clusters, whereas the K-means assumes that the number of clusters is known a priori.

5.1.2 SUPERVISED CLASSIFICATION

In supervised classification, the user has a priori knowledge of the features present within a scene. The user selects the training sites, and the statistical analysis is performed on the multiband data for each class. Instead of clusters in unsupervised classification, the supervised classification approach uses pixels in the training sets to develop appropriate discriminant functions that distinguish each class. All pixels in the image lying outside training sites are then compared with the class discriminants and assigned to the class they are closest to. Pixels in a scene that do not match any of the class groupings will remain unclassified.

Some of the common supervised classification techniques include the minimum-distance-to-means, parallelopiped classifier, maximum likelihood classifier, and others. The eCognition nearest neighbor classifier employs the minimum-distance-to-means classifier for classification. Let us briefly review some of these techniques.

Minimum-distance-to-means: The minimum distance classifier sets up clusters in multidimensional space, each defining a distinct class. Each pixel within the image is then assigned to that class it is closest to. This type of classifier determines the mean value of each class in each band. It then assigns unknown pixels to the class whose means are most similar to the value of the unknown pixel.

Parallelopiped classifier: One of the simplest supervised classifiers is the parallelopiped method. This classifier works by delineating the boundaries of a training class using straight lines. In case of two-band imagery, the boundaries will look like a series of rectangles. After the boundaries have been set, unknown pixels are assigned to a class if they fall within the boundaries of the class. If the unknown pixel does not fall within the boundary of any class, it is classified as unknown. This method is computationally efficient and attempts to capture the boundaries of each class. However, using straight lines to delineate the classes limits the method's effectiveness. Also, having pixels classified as unknown may be undesirable for some applications.

Maximum likelihood classifier (MLC): The most powerful classifier in common use is the maximum likelihood classifier. Based on statistics mean, variance/covariance, a Bayesian probability function is calculated from the inputs for classes established from training sites. Each pixel is then judged as to the class to which it most probably belongs.

5.1.3 Rule-Based Classification

Rule-based classification evolved from the expert systems domain where a given occurrence can be explained by a set of rules and instances. Each instance results in a decision, and the associated rule set is comprised of a series of logical steps that are built on an existing set of variables to explain the occurrence. ERDAS Imagine from Leica Geo Systems has a built-in module called *spatial modeler* for developing the rule bases for image classification. If the user is familiar with the spectral behavior of the objects and the associated phenomenology, a rule base can be developed that can capture the knowledge of the user analogous to the knowledge engineer in expert systems. Following is a simple example of thematic classification of water and vegetation using spectral reflectance values. The rule bases similar to the rules shown here can be created to extract water and vegetation features:

If NIR < 20% and Blue < 4%, then pixels belong to water.

To classify vegetation, we can use the vegetation index NDVI. The associated rule for classifying vegetation will look similar to the following rule:

If NDVI > 0.4, then the pixel belongs to vegetation.

When the user is not familiar with spectral behavior and phenomenology of various features, there are several data mining techniques available to understand the relationship between a specific thematic class and independent variables, such as spectral bands, derived information layers, such as NDVI, tassel cap, indices, and any ancillary data layers, such as elevation, slope, and aspect. Histogram plots, statistical analyses such as Duncan classification, and multidimensional plots can help the user to select appropriate inputs to feed into these data mining techniques.

In later sections we will discuss statistical analyses, such as CART and decision trees, that can be used to identify appropriate variables for classification.

Let us redo Example 4.5 and Example 4.6 from Chapter 4, but this time, use a rule-based classification approach rather than a nearest neighbor supervised classification approach. We will also explore some of the functionality with the eCognition software that will allow us to minimize number of objects and export thematic maps as vectors as well as raster.

EXAMPLE 5.1: RULE-BASED CLASSIFICATION

Step 1: Create a new project and import example5a.tif.

Step 2: Create two classes in the Class Hierarchy window, Water and Vegetation.

When creating the classes, assign the color blue to water and green to vegetation. We will be able to export the color tables for the final classified image.

Step 3: Create a rule base for water.

Using eCognition, see the range of near-infrared (NIR) values (Band 1 in this scenario) of various features. Navigate to Feature View → Object Features → Mean Band1. Move the cursor on water features in the main display window, and the digital number (DN) values of NIR band for water will be displayed. Notice that the values range roughly from 0 to 40 DNs. eCognition has the functionality to display only selected ranges in the display window. To accomplish this, check the check box at the bottom of the Feature View window, and type in 0 for left border value and 40 for right border value. Lower values that are close to 0 are displayed in blue color and higher values in green, as shown in Figure 5.1.

This feature is very useful in identifying value ranges for various features and developing corresponding rule sets. Now that we know the range of DN values of water in the NIR band, let us create a rule in eCognition. Open the water class that we have created in the Class Hierarchy window and double-click on and[min] and navigate to Layer Values and Mean → Example 5a(1).tif. Add a rule as shown in Figure 5.2.

To create the rule base, we need to select a function that will encompass all the DN values between 0 and 40, including either values. Select an appropriate membership function under the Initialize option, in this case, the last icon in the second row. Now, we need to define the range of values for this membership function. Type in value 0 for left border and 40 for right border. The Entire Range of Values option should display the possible values you can type in, and in this case, as this is 8-bit data, we can type in values between 0 and 255. We can leave the unit option to No Unit. We will be primarily using the unit in examples where we are using distance measures to create a rule.

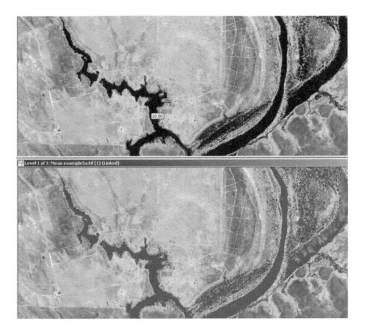

FIGURE 5.1 (Color figure follows p. 82.) Displaying ranges of values in Feature View window.

Step 4: Let us now create a custom feature called NDVI. Click on the icon Create Custom Features and create the rule as shown in Figure 5.3.

For this exercise, we will use the arithmetic functions. The default-naming convention for custom features in eCognition is arithmetic 1, arithmetic 2, and so on, but the user can modify the feature name. Let us type in NDVI in the feature name. NDVI is computed as follows:

$$NDVI = (NIR - Red)/(NIR + Red)$$

Using the calculator below, type in the equation as shown in Figure 5.3. Notice that in the right-hand window, you have access to all object features. As this customized features window also has functions to perform trigonometric operations including inverse functions, we can now calculate complex formulae such as spectral angle mapper (SAM) inside eCognition. Also, notice that the option Degrees and Radians does the conversion and outputs the result directly in the units desired by the end user. One common application where these trigonometric functions come in handy is during the conversion of DNs into radiance and correcting for solar zenith angle.

Now, display the values of NDVI. In Feature View window, all custom features are listed under the Custom Features option on the top. Navigate to Customized View, and select NDVI. Similar to the way we displayed NIR values before, these steps will now display NDVI values of objects in the main display window. Move the cursor in the display window to find the range of NDVI for vegetation features.

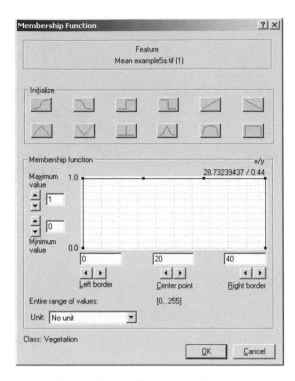

FIGURE 5.2 Creating a rule base for classifying water features.

You will notice that NDVI > 0 is vegetation. You can apply the rule as shown in Figure 5.4.

We will follow the same steps we have used for creating the rule base of water. Open the Vegetation class in the Class Hierarchy window and double-click on and[min] option. Navigate to Custom Features → NDVI and select the feature. To set up the membership function that reflects the rule that NDVI > 0 is vegetation, select the appropriate membership function shown in Figure 5.4. The shape of the selected function results in crisp classification where values greater than center point are classified as vegetation, and values less than the center point are not. Type in the value 0 for the center point.

As typical NDVI values range from 0–1, we could have used the same function we used for water and typed in 0 and 1 for lower and right borders, respectively.

Step 5: Classify.

Step 6: Display object boundaries.

Step 7: Let us now merge contiguous objects with the same class into one object. To accomplish this, click on the Structure tab in Class Hierarchy and drag the features Water, Vegetation, into a New Structure Group as shown in Figure 5.5.

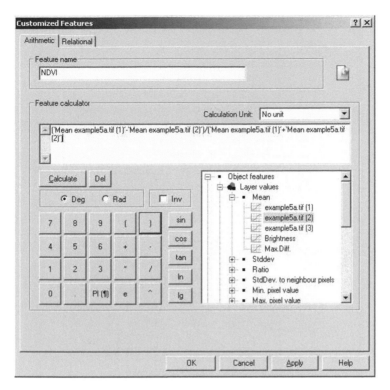

FIGURE 5.3 Creating the custom feature NDVI, within eCognition.

Note: If you want to make changes to the classes, you have to select the Inheritance tab in the Class Hierarchy window.

Step 8: Go to Level 3, and select the icon Classification Based Segmentation, or use the menu option Image Objects → Classification Based Segmentation.

Step 9: Select the option Merge Image Objects in Selected Level as shown in Figure 5.6.

Step 10: Display object boundaries.

As you will notice, all the contiguous water objects are now merged into large polygons. Now, let us export the classification into a geographic information system (GIS)-ready format, for use in GIS software.

Step 11: Let us export the current classification as a raster.

Select Export → Classification from the menu option.

As you will notice, multiple classifications on different levels can be exported separately. If desired, you could only export water or vegetation layers or the combined layer. The default format for raster output is GeoTIFF, but eCognition lets you export to other image formats such as .img and .pix formats. The export

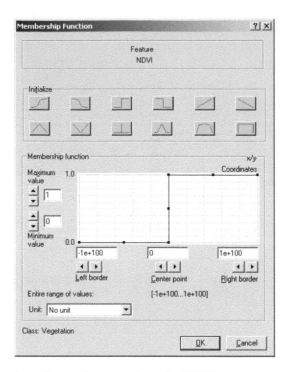

FIGURE 5.4 Creating rule base for vegetation using NDVI.

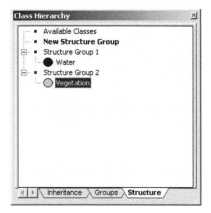

FIGURE 5.5 Merging contiguous objects of same class.

option in eCognition creates two files: a GeoTIFF file and an associated comma-separated-values ascii file, .csv file, which contains class number and associated RGB values used for class colors.

Step 12: Now let us export the classification as vectors.

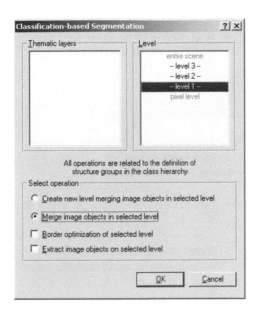

FIGURE 5.6 Merging objects.

Before we do that, let us create vector polygons by clicking on the Create/Modify Polygons icon. This will give an option to set thresholds for polygons and also remove any slivers. Now, select the menu option Export and Image Objects. By selecting the option Classification Based and Class Color, the output shapefile will contain database columns for Best Classification, Associated Classification Value, and columns for RGB, selected for various classes. You have two options for exporting polygons: (1) smoothed and (2) raster. In general, smoothed rasters are aesthetically pleasing as compared to raster polygons. We will look into the shapefile attributes of the exported polygons in other examples. Also, notice that you can export only selected classes in a raster format.

There is one more option for exporting vectors. Let us select Object Shapes from the export menu. This export option is very useful and allows exporting the polygons as point, line, or polygon. Also, unlike the Export Image Objects option, you can export only selected classes and associated attributes. If you are extracting linear features such as streams, the export line option automatically creates centerlines of polygons. Use the export point option in case of exporting features such as building centroids. Explore different export options of Object Shapes.

Step 13: The last export option is Export Current View, which is the option for individual feature layers to be displayed in the main window.

To display only water features, navigate to Feature View, Classified As, and select Water. The main window will display a binary raster with water features as 1 and the other values as 0.

Now you can export individual classes as JPEGs or another bitmap format for display or to use in other GIS software. Also, this option exports the GeoTIFF format, which retains geographic information of the scene.

Step 14: Save project.

In this example we have learned to:

• Create rule bases for classification
• Merge contiguous objects with same class
• Export classification in various formats

5.1.4 CLASSIFICATION AND REGRESSION TREES (CART) AND DECISION TREES

Tree-based modeling is an exploratory data mining technique for uncovering structure in large data sets. This technique is useful for the following applications:

• Constructing and evaluating multivariate predictive models
• Screening variables
• Summarizing large multivariate data sets
• Assessing the adequacy of linear models

Tree-based models are useful for both classification as well as regression problems. The problem domain consists of a set of classification or predictor variables (X_i) and a dependent variable (Y). In image analysis, typically, the (X_i) variables may be a combination of spectral bands, vegetation indices, ancillary data sets such as elevation and Y, a quantitative or a qualitative variable such as a feature class. In classification trees, the dependent variable is categorical, whereas in regression trees, the dependent variable is quantitative. Regression trees parallel regression analysis of variance (ANOVA) modeling, whereas classification trees parallel discriminant analysis.

Classification and regression trees (CART) are computed on the basis of a search algorithm (an iterative procedure) that asks this question: What dichotomous split on which predictor variable will maximally improve the predictability of the dependent variable? The search module carries out sequential binary splits according to a local optimization criterion that varies with the measurement scale of the dependent variable. There are several commercial software available for CART analysis, including the decision trees in ENVI software from RSI (www.rsi.com), See5 and Cubist from RuleQuest (www.rulequest.com), CART 5.0 from Salford Systems (www.salfordsystems.com), ERDAS Imagine, and other sources.

eCognition has a data mining functionality called Feature Space Optimization that will allow us to identify the best parameters to use for classification, based on training sets. One of the important aspects of data mining techniques to keep in mind is that no correlation is assumed between the input variables. The user needs to be cognizant about the correlation and needs to ensure that the input variables are not highly correlated. Also, the more the number of variables that are used, the longer is the processing time required for data mining. Keeping this fact in mind, let us investigate the Feature Space Optimization functionality in eCognition.

EXAMPLE 5.2: EXPLORING FEATURE SPACE OPTIMIZATION

Step 1: Import example8.tif.

Step 2: Segment the image at a scale parameter of 10.

Step 3: Create a class hierarchy with four classes: (1) Trees, (2) Golf Course, (3) Bare Soil, and (4) Built-up.

Step 4: Using sample editor, select training sets for all four classes.

Step 5: Select the Feature Space Optimization icon or use the Menu option and navigate to Tools → Feature Space Optimization and the screen shown in Figure 5.7 will appear.

By default, all the classes are selected for feature space optimization. You can click on the Select Classes button to change the selection of features to run the data mining on. In the right-hand window, eCognition offers over 150 features for optimization. I recommend that the user select appropriate features, based on classes of interest. Click on Select Features to choose from a wide range of features available. For this exercise, let us select the features shown in Figure 5.8.

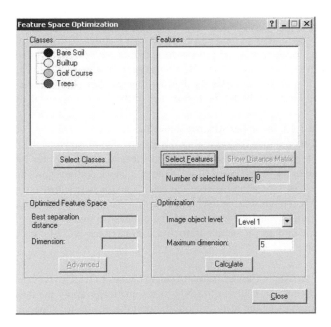

FIGURE 5.7 Feature Space Optimization interface.

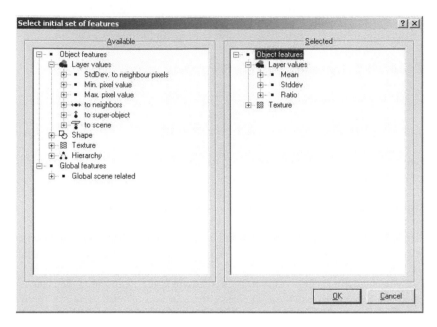

FIGURE 5.8 Features selection menu for data mining.

As shown in Figure 5.9, let us select 39 features.

Now to find the optimum features for identifying the four classes, click on the Calculate button to perform feature space optimization. The Maximum Dimension option is the number of iterations to select best features. The higher the number of iterations, the longer is the process to compute the optimum variables. The two options in the left bottom corner, Best Separation Distance and Dimension, describe the best feature space to classify the four classes after computation. Click on the Advanced button in the left bottom corner of the window to see the best fit.

In Figure 5.10, the upper window shows the results list and lists each dimension and the best features for separating the classes. Scroll down to the bottom of the window to see different dimensions and the associated features best suited for separation of the four selected classes. The plot on the bottom shows that Dimension 5 has the best separation between the classes and is best suited for separating the classes. I recommend using the iteration with peak distance measure for classification, keeping in mind the variables of that dimension are not highly correlated. Try a different number of iterations to see the variables picked by eCognition for classification.

Click on the Apply Classes button to select classes to which you want to apply the rules and click on Apply to Std NN to automatically apply the nearest neighbor classification using the features from the best dimension for supervised classification. As discussed in the previous chapter, you can use nearest neighbor (NN) classification to modify variables for each different class based on the feature space

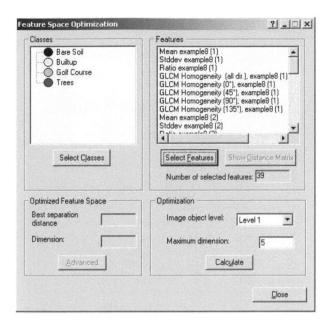

FIGURE 5.9 Features selected for data mining.

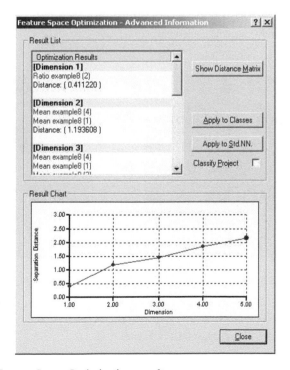

FIGURE 5.10 Feature Space Optimization results.

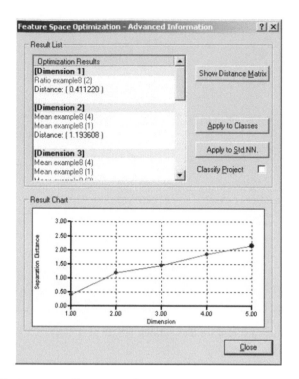

FIGURE 5.11 Feature space distance matrix.

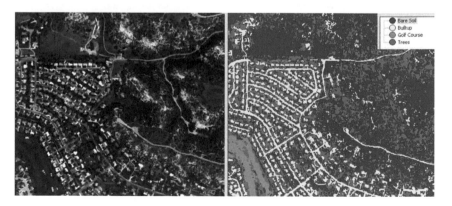

FIGURE 5.12 (Color figure follows p. 82.) Classification based on results from feature space optimization.

optimization results. Click on Show Distance Matrix to see distance separation between various classes as shown in Figure 5.11.

Step 6: Select the option Classify Project to perform classification of the image and your result should look similar to Figure 5.12.

In this example, we have learned to:

- Explore Feature Space Optimization
- Apply selected rules from Feature Space Optimization for nearest neighbor classification

5.1.5 NEURAL NETS AND FUZZY LOGIC CLASSIFICATION

5.1.5.1 Neural Network Classification

The brain is a collection of billions of interconnected neurons. Each neuron is a cell that uses biochemical reactions to receive, process, and transmit information. A neuron's dendritic tree is connected to thousands of neighboring neurons. When a neuron fires, a positive or negative charge is received by one of the dendrites. The strengths of all the received charges are added together through the processes of spatial and temporal summation. Spatial summation occurs when several weak signals are converted into a single large signal, whereas temporal summation converts a rapid series of weak pulses from one source into one large signal. If the aggregate input is greater than a threshold value, then the neuron fires, and an output signal is transmitted. The strength of the output is constant, regardless of whether the input was just above the threshold, or a hundred times as great. The output strength is unaffected by input signal strengths. Also neurons are functionally complete, which means that in addition to logic, they are also capable of storing and retrieving data from memory. A neural network can store data in two formats: permanent data (long-term memory) may be designed into the weightings of each neuron, or temporary data (short-term memory) can be actively circulated in a loop, until it is needed again. Neural networks can be explicitly programmed to perform a task by manually creating the topology and then setting the weights of each link and threshold. However, the unique strength of neural nets is their ability to program themselves.

One common technique is the multilayer perceptron (MLP) neural network, which is based on a back propagation algorithm. The MLP model uses the back propagation algorithm and consists of sets of nodes arranged in multiple layers called input, output, and hidden layers. A schematic of a three-layer MLP model is shown in Figure 5.13. Only the nodes of two different consecutive layers are connected by weights, but there is no connection among the nodes in the same layer. For all nodes, except the input layer nodes, the total input of each node is the sum of weighted outputs of the nodes in the previous layer. Each node is activated with the input to the node and the activation function of the node.

The input and output of the node I in an MLP model according to the back propagation algorithm is:

Input: $X_i = \Sigma W_{ij} O_j + b_i$ (1)
Output $= O_i = f(x_i)$ (5.3)

The back propagation algorithm is designed to reduce errors between the actual output and the desired output of the network in a gradient descent manner. The summed squared error (SSE) is defined as:

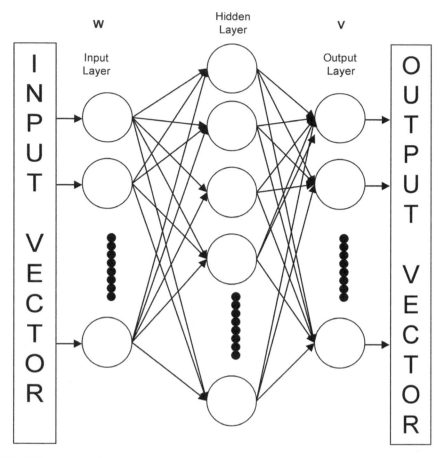

FIGURE 5.13 Multiperceptron neural network.

$$SSE = \Sigma_p \, \Sigma_I \, (O_{pi} - t_{pi})^2 \qquad (5.4)$$

where p indexes all the training patterns, and i indexes the output nodes of the network. O_{pi} and t_{pi} denote the actual output and the desired output of node i, respectively, when the input vector p is applied to the network.

A set of representative input and output patterns is selected to train the network. The connection weight W_{ij} is adjusted when each pattern is presented. All the patterns are repeatedly presented to the network until the SSE function is minimized. Neural-networks-based classification methods are not dependent on statistical distributions of spectral response of different classes in the test and training data sets. Further, neural-networks-based classification methods are not dependent upon the choice of training data set sizes and spatial characteristics. Neural net classification methods are accurate enough even with small training data sets.

Neural nets nodes are similar to the CART data mining techniques we have discussed before. For a given output, the hidden layers use a combination of input layers and assign weights to derive the expected outcome. I recommend "Neil's Neural Nets" (http://vv.carleton.ca/~neil/neural/) to gain an understanding of neural nets technology.

5.1.5.2 Fuzzy Classification

Let us now look at fuzzy classification procedures.

Fuzzy classification is a probabilistic approach and a powerful classification technique that uses expert system rules for classification and is best suited to exploit the spectral family of signatures for a given class and spectral overlap between classes, due to limited spectral resolution. For a thematic map with n classes, fuzzy classification includes n-dimensional membership degrees, which describe the likelihood of class assignment of an object *obj* to the n considered thematic classes.

$$f_{class, obj} = [\mu(obj), \mu_{class\_1}(obj), \dots \mu(obj)]$$

$$class\_1 \quad class\_n$$

Techniques such as K-means supervised classification are crisp classification techniques that result in a binary assignment of an object to the available set of thematic class, whose membership degree is the highest. In crisp classification, every object is assigned to a specific class, in spite of the weak correlation between object properties with thematic class attributes. On the other hand, the fuzzy rule set contains information about the overall reliability, stability, and class combination of all the potential classes the object can belong to, including an option for an unclassified object that does not meet the membership function requirements of all the n classes within the image.

Fuzzy classification requires a complete fuzzy system, consisting of fuzzification of input variables that will form fuzzy sets, fuzzy logic combinations of these fuzzy sets for defining a class, and defuzzification of the fuzzy classification results to get the common crisp classification for thematic classification. Fuzzy logic is used to quantify uncertain/qualitative statements into a range of data values that replace the two boolean logical statements "true" and "false" by a continuous data range 0 … 1. The value 0 represents the qualitative value of "false" and 1 represents "true." All the values between 0 and 1 represent a transition between true and false. Fuzzy logic can take into account imprecise human thinking and can implement rules that mimic the human interpretation process. In eCognition, the fuzzy sets are defined by membership functions.

Fuzzy classification process consists of three steps: (1) fuzzification of thematic class descriptions, (2) fuzzy rule base creation, and (3) defuzzification.

5.1.5.2.1 Fuzzification

Fuzzification describes the transition from a crisp system to a fuzzy system. Objects are given a membership degree (that ranges from 0 to 1) for more than one thematic class in the scene. An example would be where a dark object, which is water, can be assigned a membership degree, $\mu(obji)$ of 0.99 for water, $\mu(obji)$ of 0.95 to the shadow class and, finally, a $\mu(obji)$ of 0.78 so that it can also be an asphalt roof top.

In a scene with n number of thematic classes, each class can have its own fuzzy rule base, which is a collection of rules using various object parameters: spectral, spatial, morphological, and textural. Further, each rule can have a different shape of the membership function that will result in object classification ranging from "full member" to "not a member." For a given object parameter, all thematic classes that have a membership value higher than 0 belong to a fuzzy set of the object. In general, the broader the membership function, the more vague the underlying concept. An example would be saying that all the objects with DN in the blue band less than 60 belong to water. Other thematic classes such as shadows, asphalt roads, and other dark objects will all be assigned a membership value for this parameter. The lower the membership value, the more uncertain is the assignment of a certain value to the set. Let us investigate how to develop these rule bases.

5.1.5.2.2 Fuzzy Rule Base

A fuzzy rule base is a combination of fuzzy rules, which combine different fuzzy sets that make up a description of a thematic class. The simplest fuzzy rules are dependent on only one fuzzy set. Fuzzy rules are "if–then" rules. If a condition is fulfilled, an action takes place. Let us look at spectral curves in the Figure 5.14.

The following rule can be constructed to classify water:

If blue reflectance < 5%, then image object is water

This rule is applicable only for clear and deep water bodies, and if the scene contained spectral signatures as shown above, the result would be a crisp classification of the object being assigned to water class. In reality, water spectral signatures are not unique lines as shown above but are rather a family or collection of signatures. The blue band reflectance values can range from 0% to 10% based on whether the water is clear or turbid, water depth is deep versus shallow, if the conditions at the time of capture are calm or windy, and other factors. We modify the previous rule as follows:

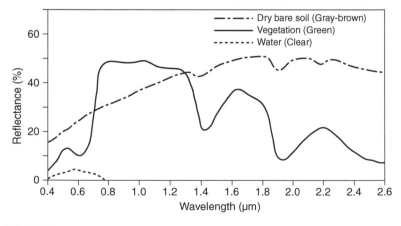

FIGURE 5.14 Spectral reflectance of various features.

If blue reflectance < 10%, then image object is water

The same rule will result in membership values for all three classes, similar to the following values:

$$\mu_{water}(obj) = 0.97$$

$$\mu_{soil}(obj) = 0.7$$

$$\mu_{vegetation}(obj) = 0.0$$

Although water has the highest membership value using the rule above, the possibility is high that the object can also belong to soil. We need additional rules that exploit low reflectance values of water in other bands to defuzzify object classification. So we can add another rule as follows to classify water:

If blue reflectance values < 10% and
If NIR reflectance values < 8%, then the object is possibly water

This fuzzy set, a combination of fuzzy rules, might result in membership values as follows:

$$\mu_{water}(obj) = 0.975$$

$$\mu_{soil}(obj) = 0.32$$

$$\mu_{vegetation}(obj) = 0.0$$

You will notice that the membership value of soil has decreased significantly, and this fuzzy set would be a good set to classify water bodies. Knowledge of spectral, spatial, contextual, and morphological parameters of a feature will help the user to better define these fuzzy-rule sets.

Fuzzy classification gives a possibility for an object to belong to a class, whereas classification based on probability provides a probability to belong to a class. Probability relies on statistics and gives information on many objects, and probability of all possible events adds up to one. This is not necessarily true with possibilities. As you will notice in the previous two results, the sum of membership values is >1. The higher the membership degree for the most possible class, the more reliable is the assignment. The bigger the difference between highest and second highest membership value, the clearer and more stable the classification. Classification stability and reliability can be calculated and visualized within eCognition as an advanced method for classification validation. Let us look at some scenarios of fuzzy classification possibilities. If the membership values for an object are all high as follows, then:

$$\mu_{water}(obj) = 0.97$$

$$\mu_{soil}(obj) = 0.92$$

$$\mu_{vegetation}(obj) = 0.905$$

The membership degrees indicate an unstable classification between these classes. The fuzzy set class definition is not sufficient to distinguish between these classes, or the spatial or spectral resolution is not suitable for this classification.

If we encounter membership values as follows:

$$\mu_{water}(obj) = 0.24$$

$$\mu_{soil}(obj) = 0.21$$

$$\mu_{vegetation}(obj) = 0.20$$

The low membership degrees indicate an unstable classification between these classes for the considered image object. However, in this case, the low membership value indicates a highly unreliable assignment. Assuming a threshold of a minimum membership degree of 0.3, no class assignment will be given in the final output.

5.1.5.2.3 Defuzzification

To produce thematic maps for standard land use and land cover (LULC) applications, the fuzzy results have to be translated back to a crisp value, which means that an object is either assigned to a particular class or is unclassified. For defuzzification of classification results, the class with the highest membership degree is chosen. Defuzzification results in a crisp classification. If the membership degree of a class is below a certain value, no classification is performed to ensure minimum reliability. As this output discards the rich measures of uncertainty of the fuzzy classification, this step should be the final step in the whole information extraction process.

$$f_{crisp} = \max \{\mu\,(obj), \mu(obj), \dots , \mu(obj)\}$$

$$class\_1 \;\; class\;2\_ \;\; class\_\,n$$

with class assignment equal to the class i with the highest members.

Let us look at a practical example using multispectral imagery to investigate the fuzzy classification.

EXAMPLE 5.3: FUZZY CLASSIFICATION

Step1: Import example9.tif.

This is a scene from a SPOT 5 sensor (www.spot.com), and the bands are arranged in the order: NIR, Red, Green, and short wave infra-red (SWIR). SPOT 5 captures multispectral data at 10 m GSD.

Step 2: Segment the image at a scale parameter of 10.

Step 3: Create a class hierarchy with five classes: Water, Built-up, Bare Soil, Trees, Parks and Lawns.

Step 4: Select samples for each of the classes using the sample editor.

Step 5: Let us look at the spectral distribution of various classes. Navigate to Tools → 2D Feature Space Plot.

The spectral window pops up. Let us display the spectral plot between NIR and Red. Select "Mean example 9(1).tif" for Y-axis and "Mean example 9(2).tif" for X-axis. You should see the plot as shown in Figure 5.15.

You will notice that water is located at the bottom left of the plot. The spectral signature of Bare Soil is entirely contained within the superset of Built-up class. Let us use a different tool to see spectral overlap of these two classes.

Step 6: Click on the Object Informtion icon to see the spectral values in all four bands.

Select the active class as Built-up and compare it with the Bare Soil class as shown in Figure 5.16.

Select various objects within the scene to see where they fall on the spectral distribution of Built-up and Bare Soil classes.

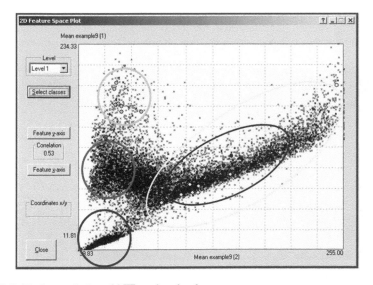

FIGURE 5.15 Spectral plot of NIR and red values.

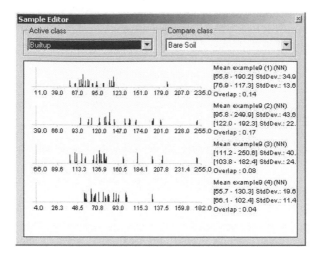

FIGURE 5.16 Comparison of spectral signatures of samples.

Step 7: Now classify the scene by adding standard nearest neighbor rules to all classes.

Step 8: Select any object classified as Built-up.

eCognition displays the object classification and membership function or associated probability associated with that classification as shown in Figure 5.17.

Now, let us investigate the confusion between various classes. Click on the Classification tab in the Class Hierarchy window as shown in Figure 5.18.

Under the Classification tab in the window, you will notice that the object I selected is classified as Built-up with a highest membership function of 0.799. Also, notice that it is very close to Bare Soil and has a membership value of 0.783 with the possibility that it belongs to soil. The alternative assignments show the object

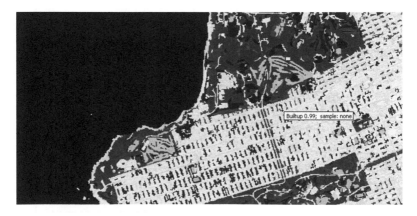

FIGURE 5.17 Displaying probability of object classification.

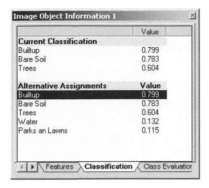

FIGURE 5.18 Object Information window.

membership functions to various classes. This is a valuable tool for quality control, and you can set up rules to identify objects that have two or more classes with high membership values, indicating the classification of an object is unstable.

eCognition also allows us to set a minimum membership value to eliminate objects with low possibility being assigned to an object. Navigate to Classification → Advanced Settings → Edit Minimum Membership Value.

This function is useful, especially if you are exporting the classification and performing spatial queries in a GIS.

Step 9: Let us identify all the objects that are classified as either Bare Soil or Built-up and have high membership values to both classes. Create a new class called Spectral Mix and add the following rules shown in Figure 5.19.

Step 10: Let us perform classification with class related features.

Figure 5.20 shows the objects that have membership functions > 0.7 for both Bare Soil and Built-up.

In this example we have learned to:

- Create spectral plots
- Compare spectral signatures between classes
- Investigate membership functions after classification
- Develop a rule base to identify objects with mixed classes

5.2 SUPERVISED CLASSIFICATION USING MULTISPECTRAL INFORMATION

In this section, we will perform image classification using NN classification approach within eCognition. This chapter builds on the segmentation and classification steps we have done in the previous sections. Further, we will also explore the use of fuzzy

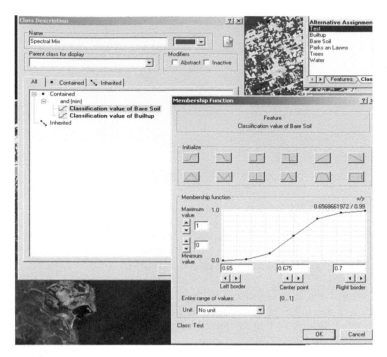

FIGURE 5.19 Creating a rule to identify objects with unstable classification.

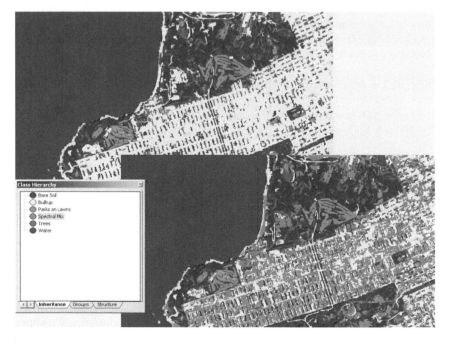

FIGURE 5.20 (Color figure follows p. 82.) Map showing objects with unstable classification.

TABLE 5.1
Spectral Features in eCognition

Object Spectral Feature	Application Tips
Mean spectral values	Vegetation has high NIR values, and water has low spectral values in all bands
Brightness	Bright objects such as metal rooftops, snow, and others have high brightness values Water bodies, shadows, asphalt, and other dark objects have low brightness values
Ratios	Blue ratio is an important tool that can be used to identify water and shadows Green and Red ratios are useful for bare soil identification
MaxDiff	Useful for impervious surface extraction
Standard deviation	Large bare soil patches tend to have low red standard deviation

logic classifiers. eCognition provides a variety of spectral features in addition to the spectral bands for thematic classification. Table 5.1 lists some of the commonly used spectral features that can be used for image analysis.

Let us do an exercise where we extract water bodies using various spectral measures.

EXAMPLE 5.4: EXTRACTION OF WATER BODIES USING SPECTRAL PARAMETERS

Step 1: Insert example10_10.tif, example10_20.tif, example10_30.tif, and example_40.tif.

These are Landsat 7 bands at 30-m resolution. In this exercise, we will use the VNIR bands of Landsat.

Step 2: Segment image at a scale parameter of 10.

Step 3: Create a class called Water.

Step 4: Now let us create a rule-based class using samples.

Step 5: Open the Sample Editor window.

Step 6: Select samples of water, making sure you select samples in shallow water as well as deep water.

Step 7: Let us create our first fuzzy classification rule for water. Right-click in the sample window and select the option Membership Functions → Compute and the window shown in Figure 5.21 will appear.

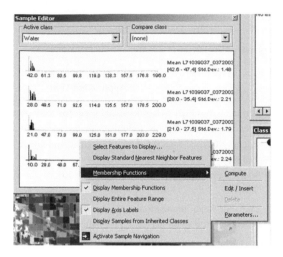

FIGURE 5.21 Computation of membership functions based on samples.

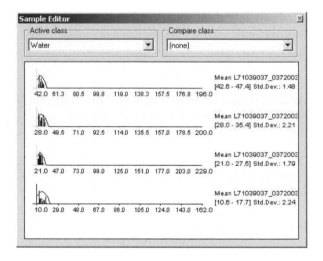

FIGURE 5.22 Membership functions derived from eCognition.

Membership functions for each band are developed automatically by eCognition. Right-click on each of the bands to develop their respective fuzzy logic rule bases as shown in Figure 5.22.

Now, let us check the rule base by double-clicking the Water class in the Feature View window. The rules, automatically generated by eCognition, should appear as shown in Figure 5.23.

Let us select the first line to see the rule created for the blue band. The screen in Figure 5.24 should appear.

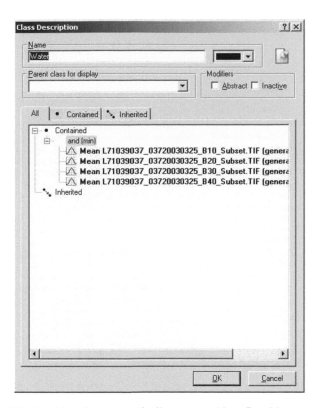

FIGURE 5.23 Membership rules automatically generated by eCognition.

This display window allows you to create various fuzzy logic rules. Let us examine various features within the Membership Function window. The top box shows the feature being used to create a fuzzy logic rule. The Initialize option in the window presents the users with 12 predefined membership functions for developing fuzzy logic rules. The membership function plot displays the range of feature values on the X-axis and associated membership range on the Y-axis. Let us develop a fuzzy rule for a water feature where we defined any values of mean blue band between 39.56 and 50.478 DNs have a possibility of being water, whereas values outside the DN range (<35.56 and >50.478) have 0 possibility/membership function to water. Objects with blue band values between 39.56 and 50.478 are given various membership functions of water based on the type of membership function selected. As shown in Figure 5.24, an object with mean DN blue value of 45.02 will have a membership value of 1.0, and an object with mean DN value 41 will have a membership value of approximately 0.6 as shown in Figure 5.25.

Similarly, an object with blue DN value of 46 will have a membership value of 0.9. Further, eCognition allows the users to customize the shape of the predefined membership functions by editing the blue connectors in the plot. It is also possible to change the range of membership functions on the Y-axis by changing the maximum and minimum values for defining the fuzzy logic rules. To get an idea of the range of values of a given object feature, the Entire Range of Values option displays

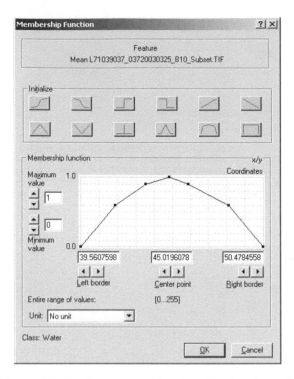

FIGURE 5.24 Membership function generated by eCognition-based samples.

the minimum and maximum values of the feature that can be used for the fuzzy logic rule. It is important to note that imagery collected at 11-bit or 12-bit radiometric resolution is often packed to 16 bits, and eCognition displays the value range from 0 to 65355 rather than the actual dynamic range of the data. The unit option within the Membership Function window displays the appropriate units for object features such as distance, shape, and area. Investigate the range of values eCognition has created for all four fuzzy logic rules, one for each band.

Step 8: Classify.

Step 9: If some of the water is still unclassified, add more samples and recompute the membership functions.

Step 10: Let us now investigate the membership value of water objects.

Select any object classified as water and click on the Class Evaluation tab in the Class Hierarchy window. This window will show the overall membership function and membership values of each of the four bands as shown in Figure 5.26.

As we used and[min] rule, which takes the least membership value of the four bands, the overall membership value is 0.660 from the green band. eCognition allows developing rules that can take the maximum of the four rules instead of the minimum. To accomplish this, click on the Inheritance tab in the class hierarchy and double-click

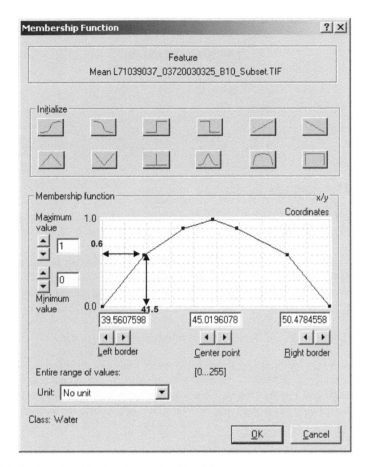

FIGURE 5.25 Membership function with eCognition.

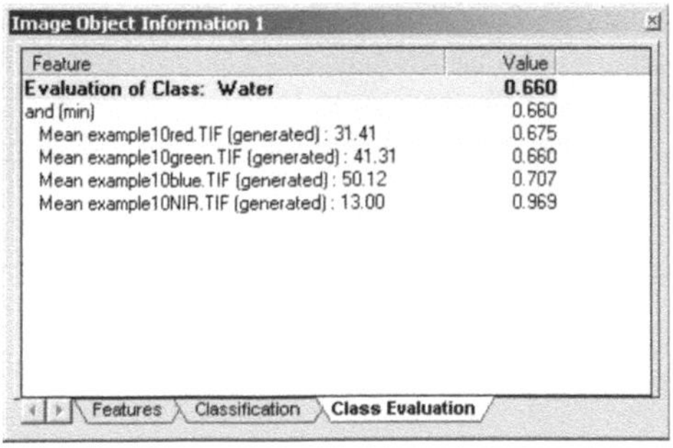

FIGURE 5.26 Class membership functions.

FIGURE 5.27 Operators for creating fuzzy logic rules.

on the Water class. Right-click on and[min] and in the following window, select Edit Expression and select the or[max] option as shown in Figure 5.27.

Rerun the classification. You will notice that any object meeting at least one of the criteria is selected as water. Selection of and[min] or or[max] option is dependent on how we define the rules. The or[max] option is used in scenarios where we can define multiple rule sets for a given class.

Step 11: Let us create a different spectral rule for classifying water.

Let us use the blue ratio for defining the water feature in this exercise. *Blue ratio* in eCognition is defined as the blue band divided by the summation of all four bands and has a value ranging from 0 to 1. Let us make the Water class in the Feature window inactive by right-clicking and unselecting the Active option as shown in Figure 5.28.

Step 12: Let us create a new class, Water-Blue Ratio.

Navigate to Feature View Window → Object Values → Ratio → Blue Band as shown in Figure 5.29.

Double-click on the blue ratio to see a display of gray scale values. You will notice the water has higher values for the blue ratio as compared to other features within the image. For this image, Water-Blue ratio values are between the values 0.35 and 0.46 as shown in Figure 5.30, and the corresponding display looks as shown in Figure 5.31.

Now let us create a membership function to classify water.

Step 13: Add the following rule to the Water-Blue Ratio class as shown in Figure 5.32.

Select the last predefined membership function from the Initialize option and type in the value 0.35 for the left border and 0.46 for the right border. Notice that you can also use the arrows to change the left and right border values. The center point is automatically computed by eCognition. As we are confident that the values between 0.35 and 0.46 are water, we will use the default values of 0 and 1 for the minimum and maximum of the membership function as shown in Figure 5.32.

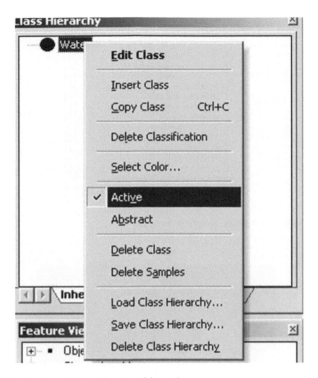

FIGURE 5.28 Activating classes in class hierarchy.

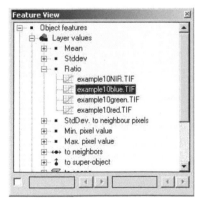

FIGURE 5.29 Using features in Feature View.

Step 14: Classify the project.

You should notice that the results are the same as using the sample approach. Although the blue ratio is good to detect water, it sometimes also captures shadows. We need additional object features to separate shadows from water.

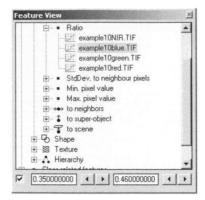

FIGURE 5.30 Selecting Feature Range in Feature View.

FIGURE 5.31 (Color figure follows p. 82.) Display of Feature Range values.

In this example we have learned to do the following:

- Create membership functions from samples
- Develop rule bases

5.3   EXPLORING THE SPATIAL DIMENSION

In this example, let us explore spatial parameters such as area, size, shape, and other spatial attributes for image analysis. eCognition has several shape parameters that can be used for feature classification. Table 5.2 lists some of the spatial features.

The following example demonstrates the use of spatial parameters along with spectral parameters to classify various water bodies like ocean, lakes, and ponds.

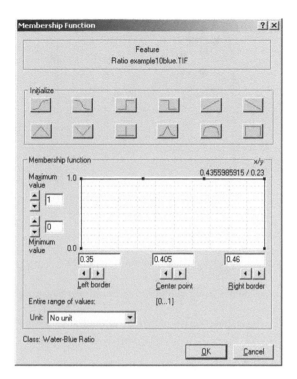

FIGURE 5.32 Creation of Membership Function.

TABLE 5.2
Spatial Features in eCognition

Spatial Feature	Application
Area	Classify water bodies into lakes or ponds, based on area
Width	Useful for road extraction
Length/width	Classify large narrow features such as streams, rivers, and man-made features like interstates
Asymmetry	Man-made objects are symmetrical
Density	Industrial and commercial areas tend to have higher density
Compactness	Man-made objects tend to be more compact than bare soil

EXAMPLE 5.5: EXTRACTION OF WATER BODIES USING SPATIAL PARAMETERS

Step 1: Import example11.tif.

Step 2: Segment the image at a scale parameter of 25.

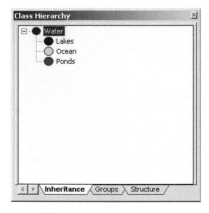

FIGURE 5.33 Creation of subclasses.

Step 3: Classify water using samples or the ratio approach.

Step 4: Merge water bodies using classification-based segmentation.

Step 5: Now let us create three classes under water to identify ocean, lakes, and ponds, based on size.

The class hierarchy should look similar to Figure 5.33.

Now, let us use shape parameter Area to differentiate the classes. By having water as a superset, the area rules to differentiate lakes, ocean, and ponds are applied to objects already classified as water.

Step 6: Let us use Feature View to display area ranges.

Step 7: Set up area rules for ocean, lakes, and ponds as shown in Figure 5.34.

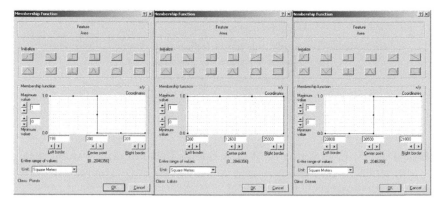

FIGURE 5.34 Membership functions for water subclasses.

FIGURE 5.35 (Color figure follows p. 82.) Classification of types of water bodies using spatial properties.

Select the appropriate predefined membership functions from the Initialize option to create the above rules.

Step 8: Classify with Class Related Features.

The output should look similar to Figure 5.35.

Step 9: Save project.

In this example we have learned to do the following:

- Create subclasses within a main class
- Use predefined membership functions for spatial attributes

Let us use a different example where we use shape parameters to extract roads.

EXAMPLE 5.6: EXTRACTION OF ROADS USING SPECTRAL PARAMETERS

Step 1: Import dq5014sw.ecw.

Step 2: Segment the image at a scale parameter of 50 and a shape parameter of 0.5.

Step 3: Create a class called Road and add rules for the following shape parameters as shown in Figure 5.36.

Use the Feature View window to figure out the values for the three parameters.

Step 4: Classify.

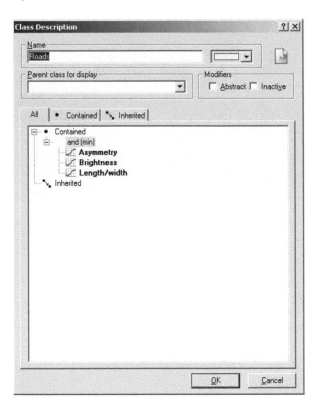

FIGURE 5.36 Feature selection for road extraction.

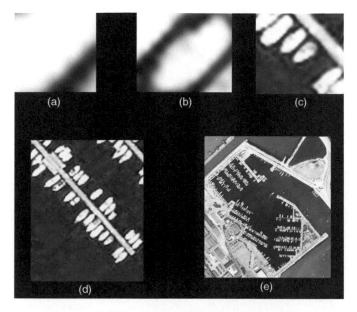

FIGURE 1.1 Representation of a marina at various zoom levels.

FIGURE 1.2 Segmented image of a marina.

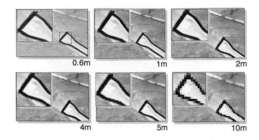

FIGURE 2.1 Spatial resolution comparison.

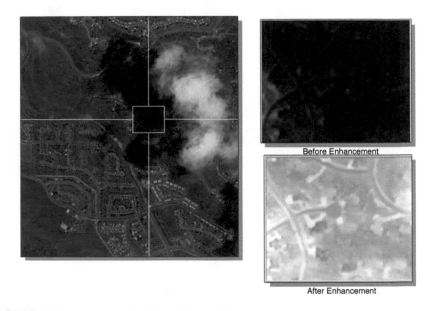

FIGURE 2.4 Importance of radiometric resolution.

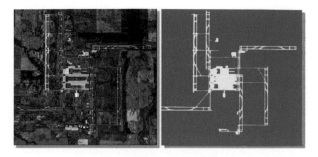

FIGURE 3.1 Supervised runway classification based on objects. In yellow are the extracted airport features.

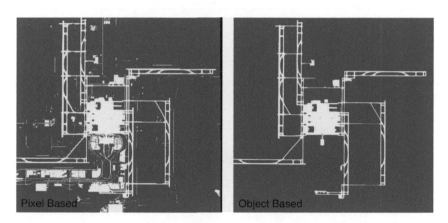

FIGURE 3.2 Comparison of pixel-based versus object-based analysis.

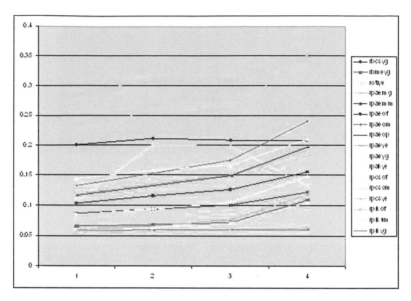

FIGURE 3.3 Spectral signatures of various roof materials and aged road pavement materials. (From UCSB.)

FIGURE 4.6 Object boundaries.

FIGURE 4.15 Training sample selection in object space.

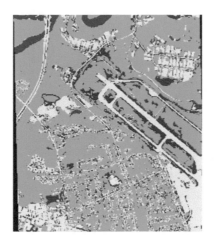

FIGURE 4.16 First object-based thematic classification results.

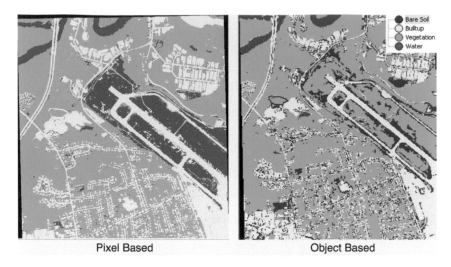

Pixel Based   Object Based

Bare Soil
Builtup
Vegetation
Water

FIGURE 4.17 Comparison of pixel-based and object-based supervised classification results.

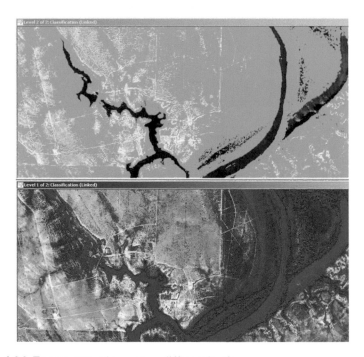

Level 2 of 2: Classification (Linked)

Level 1 of 2: Classification (Linked)

FIGURE 4.24 Feature extraction on two different levels.

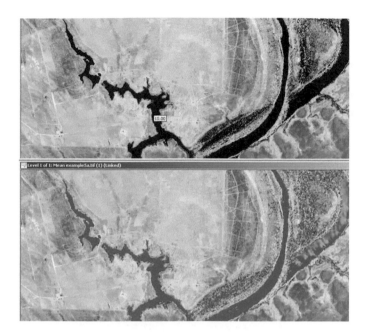

FIGURE 5.1 Displaying ranges of values in Feature View window.

FIGURE 5.12 Classification based on results from feature space optimization.

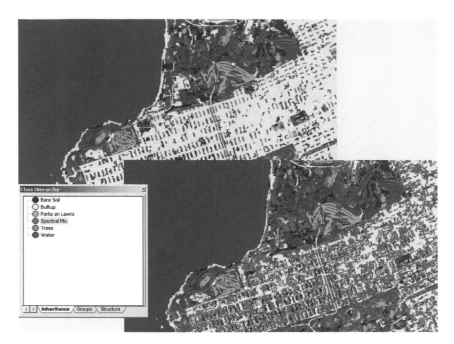

FIGURE 5.20 Map showing objects with unstable classification.

FIGURE 5.31 Display of Feature Range values.

FIGURE 5.35 Classification of types of water bodies using spatial properties.

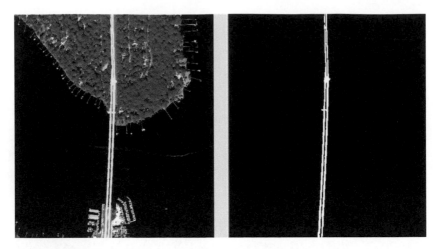

FIGURE 5.37 Feature extraction of roads.

FIGURE 5.39 Airport feature extraction.

FIGURE 5.41 Object boundaries with classification-based segmentation.

FIGURE 5.44 Classification using contextual rules.

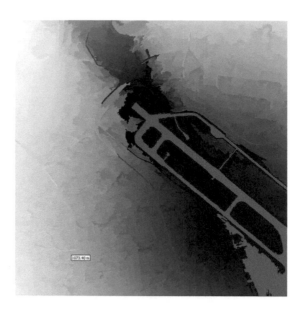

FIGURE 5.47 Distance of various objects from airport features.

FIGURE 5.49 Thematic classification of neighborhoods within proximity of the airport.

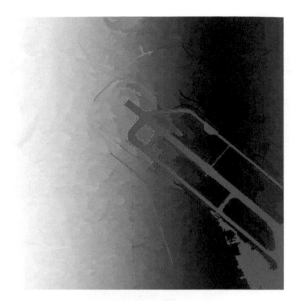

FIGURE 5.55 Display of X distances of various objects from airport.

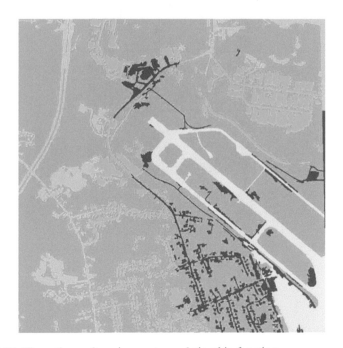

FIGURE 5.57 Thematic results using custom relationship functions.

FIGURE 5.65 Thematic map of water derived using texture.

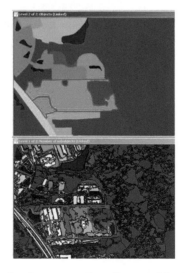

FIGURE 5.68 Display showing the preservation of parent object boundaries on child objects.

FIGURE 5.71 Object-oriented change detection results.

FIGURE 6.7 Impervious surface classification.

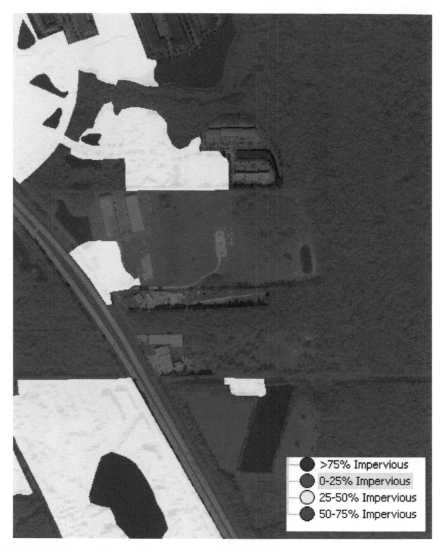

FIGURE 6.8 Impervious surfaces aggregated to parcel vectors.

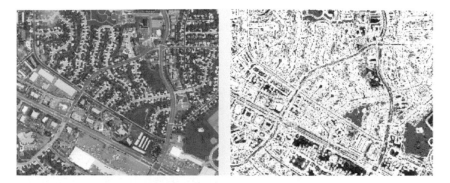

FIGURE 6.9 Image enhancement using a Sobel operator.

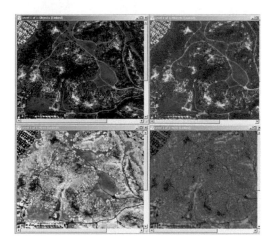

FIGURE 6.17 Normalized Difference Vegetation Index (NDVI).

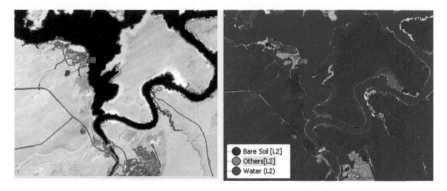

FIGURE 6.29 Thematic map of soil and water classes.

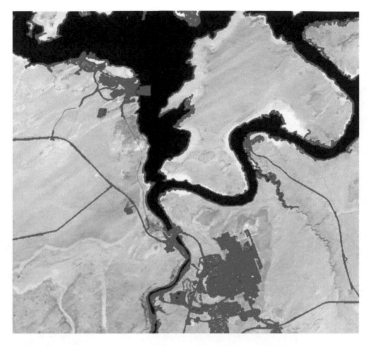

FIGURE 6.30 Object boundaries for classification-based multiresolution segmentation.

FIGURE 6.32 Thematic classification based on multiresolution segmentation.

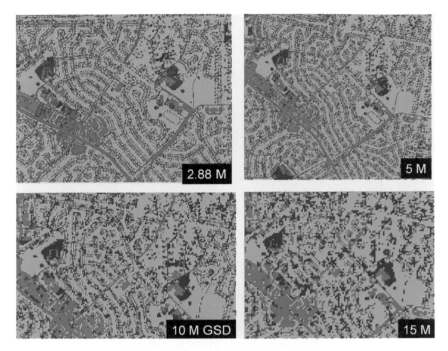

FIGURE 6.34 LULC classification at multiple GSDs.

FIGURE 6.35 Parent object boundaries preserved for lower levels of segmentation within eCognition.

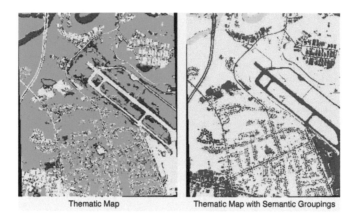

| Thematic Map | Thematic Map with Semantic Groupings |

FIGURE 6.39 Thematic map with semantic grouping.

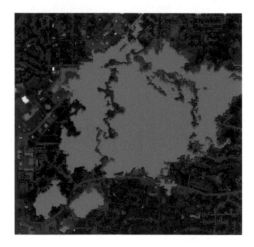

FIGURE 6.41 Semantic grouping of clouds and cloud shadows.

FIGURE 7.2 Thematic map for accuracy assessment project.

FIGURE 7.3 Classification stability display.

FIGURE 7.4 Best Classification window.

Step 5: Merge road objects and use area parameters to eliminate small features misclassified as roads.

The other option for eliminating misclassified features is the Manual Classification option in eCognition. The manual editing icons allow you to perform manual classification. The arrow icon allows you to select objects for manual classification/editing. The poly icon allows you to select objects by drawing a polygon. This option creates a polyline, and a right-click will allow closing the polygon to select the objects. The line icon allows you to draw a line and select all the objects that intersect with the line. The rectangular selection box allows you to select objects by drawing rectangles. The green icon allows editing of objects and the classify icon next to it is used for manually assigning classes to objects. To assign a class to a given object, ensure that the corresponding class is selected in the Class Hierarchy window. The icon with red and yellow objects is used to select contiguous objects for merging into one object. The chain icon will perform the actual merging of the selected objects. I recommend trying out various manual editing functions of eCognition.

For this exercise, you can select the Manual Classification icon and select objects that are misclassified. Selecting an object twice will render an object state as Unclassified. Make sure that the Roads class in the Class Hierarchy window is selected.

The final output should look like Figure 5.37.

To display the individual classes within the class hierarchy window, you can navigate to Feature View → Class-Related Features → Classified As and select roads. When you have multiple features, this is a good way of exporting individual layers as binary maps.

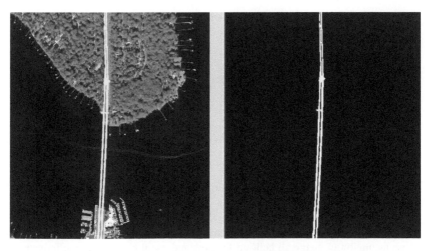

FIGURE 5.37 (Color figure follows p. 82.) Feature extraction of roads.

TABLE 5.3
Contextual Parameters in eCognition

Contextual Feature	Application
Mean difference to neighbor	Identify features that have contrast to neighbors such as roads
Std to neighbor	Residential features have large standard deviation values with neighbors
Distance to a class	This is analogous to buffering in GIS and can be used for several applications, such as identifying features within several meters of an oil pipeline, for monitoring and encroachment
Ratio to scene	Cloud detection where the cloud DN ratio has high values within the scene
Relative border to neighbors	Can be used to resolve water bodies that are misclassified as shadows Classify an island if a land object is 100% surrounded by water

In this example we have learned to effectively use the following:

• Use Feature View to identify value ranges for a given parameter
• Use shape parameters to extract rivers and streams
• Export features as lines

5.4   USING CONTEXTUAL INFORMATION

In this section, we will explore using contextual information for classifying objects. The contextual properties that can be used include distance to neighbors, relative border to certain features, and others. Table 5.3 shows some of the commonly used contextual parameters.

EXAMPLE 5.7: MAPPING NEIGHBORHOODS WITHIN PROXIMITY TO AIRPORT

Step 1: Import example12.tif.

Step 2: Segment images at a scale parameter of 100.

Step 3: Let us create two custom features NDVI and Ln[Blue].

We will use these rules along with ratios to identify airport and other built-up features.

Step 4: Let us create two classes: (1) Airport and (2) Not Airport. Use the rules in Figure 5.38 to classify airport.

Use NDVI and Ratios features to identify airport features. Use classification-based segmentation to merge all built-up features and use the area rule to eliminate any features that are misclassified. Also, use classification-based segmentation to merge

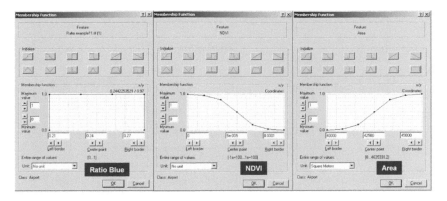

FIGURE 5.38 Rule set for airport proximity computation.

FIGURE 5.39 (Color figure follows p. 82.) Airport feature extraction.

Not Airport class features that are contiguous to each other. Your output should be similar to Figure 5.39.

For using distance measures, it is important to note that eCognition manages memory based on number of objects. The more the number of objects, the longer is the time to compute distance measures within eCognition.

Also, eCognition allows you to create a single layer with different scale parameters. Let us create a new level with objects at a finer resolution than Level 1 for objects that belong to Not Airport class. Make sure you have Not Airport class in the Structure window of the Class Hierarchy window. Let us create a sublevel at the scale parameter 25, and make sure the Classification Based check box is on as shown in Figure 5.40.

Display the object boundaries and it should look as in Figure 5.41. Notice that the airport is still one large object as compared to Not Airport features.

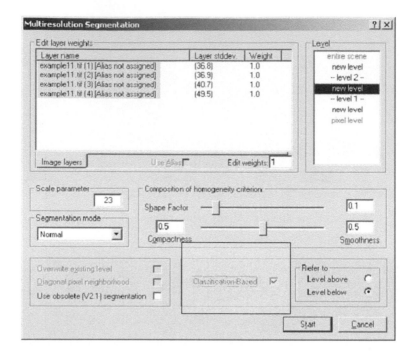

FIGURE 5.40 Classification-based segmentation.

Now let us create a class hierarchy called Level 2 and drag Airport and Not Airport under Level 2. Create a new class called Level 1 and Level 2 subclasses Airport-Mask and Built-up. Use the logic in Figure 5.42 for Airport-Mask by using Super Relationship feature.

Create a class called Built-up and use the following object features in Figure 5.43 to extract the houses, buildings, and other built-up features. Add a rule to exclude Airport features from the Built-up class. The rule set should look as in Figure 5.43.

Classify all built-up features and your output should look as shown in Figure 5.44.

Step 5: Let us classify built-up features into three classes based on distance from airport features.

Let us create three classes, within distance of $1/2$ km, 1 km, and >1 km. The final class hierarchy should look similar to Figure 5.45.

Now, let us look at the distance metric in Feature View. Navigate to Distance to Objects in class related features as shown in Figure 5.46.

Select Airport mask and the Display menu shows distances of various features to airport as in Figure 5.47.

FIGURE 5.41 (Color figure follows p. 82.) Object boundaries with classification-based segmentation.

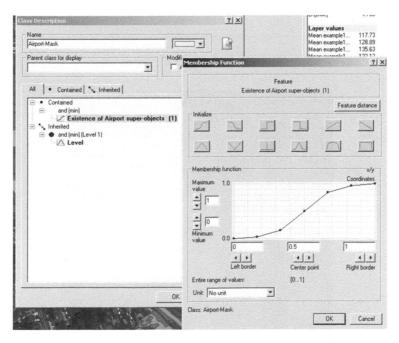

FIGURE 5.42 Using superobject relationships.

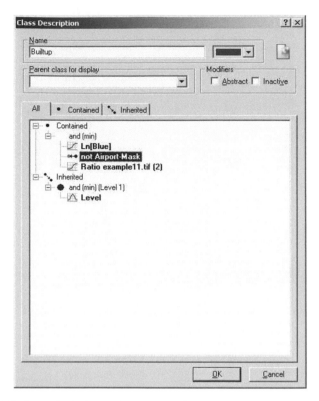

FIGURE 5.43 Rule set for built-up feature extraction.

FIGURE 5.44 (Color figure follows p. 82.) Classification using contextual rules.

FIGURE 5.45 Class hierarchy for features within various distances to an airport.

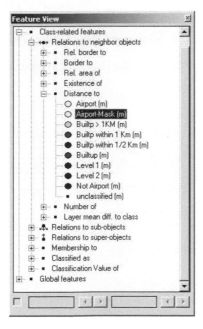

FIGURE 5.46 Distance measures to features in Feature View window.

Now let us add the rule for Built-up class that is within 0.5 km from the airport. Use the rule shown in Figure 5.48.

You can create similar rules for classes with distance to airport < 1 km and > 1 km, respectively.

Step 6: Classify using Class Related Parameters and the final thematic map should look as in Figure 5.49.

In this example we have learned to use contextual measures for image classification.

FIGURE 5.47 (Color figure follows p. 82.) Distance of various objects from airport features.

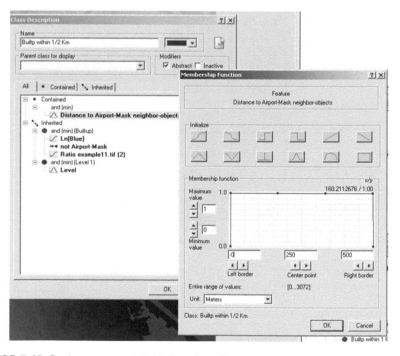

FIGURE 5.48 Setting up a spatial rule based on distance.

FIGURE 5.49 (Color figure follows p. 82.) Thematic classification of neighborhoods within proximity of the airport.

The distance measure in Feature View uses combined distances in X and Y dimensions. Let us redo the example using the logic that the distance measures can be either in X dimension or Y dimension.

EXAMPLE 5.8: MAPPING NEIGHBORHOODS WITHIN PROXIMITY TO AIRPORT USING CUSTOMIZED FEATURES

Step 1: Open project example5_7.

Step 2: Save it as example5_7.

Step 3: Delete the class rules in classes under Built-up.

Step 4: Let us create rules that will compute the distance of all objects with respect to Airport.

Click on the Custom Features window and click on the Relational Tab as shown in Figure 5.50.

Now, let us create a rule for computing X distance of objects from Airport features. Select Absolute Mean Difference by clicking on Mean Absolute Difference window as shown in Figure 5.51.

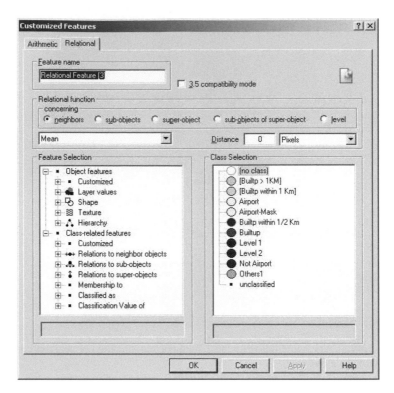

FIGURE 5.50 Creating custom relational features.

As we are primarily interested in distance of features and not whether they are to the left or right of Airport mask, we selected Mean Absolute Difference. For certain applications where we are interested, in the objects right or left of the target, select Mean Difference Parameter, which will show positive and negative values. Also, as we are interested in Neighbors to Airport in the same segmentation level, we will use the default relationship function of Neighbors.

Let us select the center X position of the Airport mask as shown in Figure 5.52.

Now, let us compute a distance of 5000 m or 5 km and ensure the units are in meters. Double-click to select the Airport Mask class and the final window should look as shown in Figure 5.53.

Step 5: Create a similar rule for Y distance.

Step 6: Let us look at the values of X ranges from Airport mask. Navigate to Class Related Features → Custom Features → X-Mean in feature view as shown in Figure 5.54.

Display the values of X Mean and move the cursor in the window to see the values as shown in Figure 5.55.

Step 7: Create rules for the three classes and Figure 5.56 shows an example of creating Built-up class within 500 m of airport features.

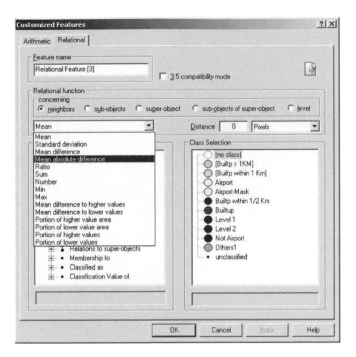

FIGURE 5.51 Creating X dimension relationship function.

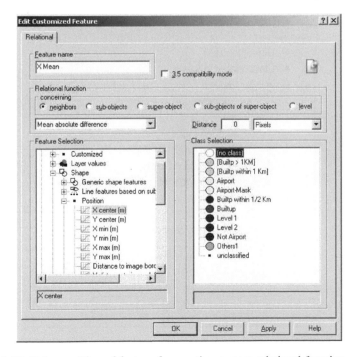

FIGURE 5.52 Using position of feature for creating custom relational function.

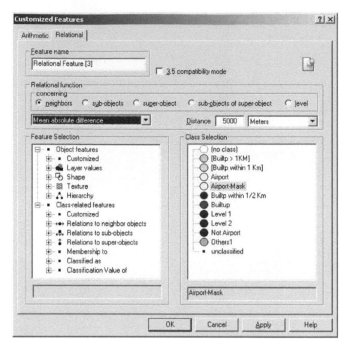

FIGURE 5.53 Using distance units for custom relationship function.

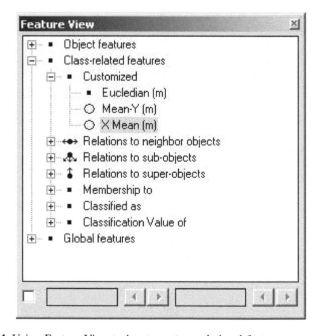

FIGURE 5.54 Using Feature View to locate custom relational features.

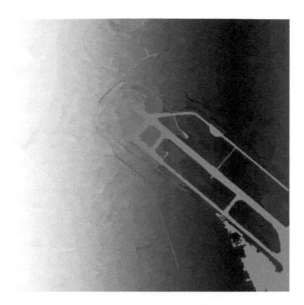

FIGURE 5.55 (Color figure follows p. 82.) Display of X distances of various objects from airport.

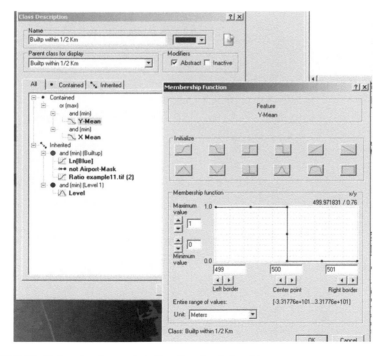

FIGURE 5.56 Rule set using custom functional relationships.

Notice that we have used or[max] option to identify objects that are within a certain distance to the airport either in X or Y direction.

Step 8: Classify and your results should look as shown in Figure 5.57.

In this example we have learned to do the following:

- Use contextual parameter of distance to a thematic class to extract features
- Use contextual parameter, Border Relationship to Neighbor Objects
- Create custom functional relationship functions for neighboring pixels

Let us do another example that uses contextual relationship for classifying an island.

EXAMPLE 5.9 CLASSIFYING AN ISLAND USING CONTEXTUAL RELATIONSHIPS

Step 1: Import example12b.tif.

Step 2: Segment image at a scale parameter of 250.

Step 3: Create classes, Water and Land. Use Brightness to classify Water and use the mask rule that Land is Not Water.

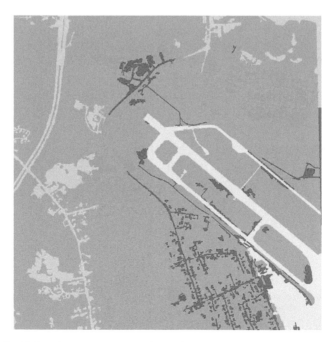

FIGURE 5.57 (Color figure follows p. 82.) Thematic results using custom relationship functions.

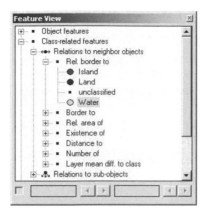

FIGURE 5.58 Using contextual object features.

FIGURE 5.59 Using contextual relationships to classify an island.

Step 4: Perform classification-based segmentation on both land and water.

Step 5: Let us create a class called Island under Land.

Step 6: Add a contextual rule that any land surrounded 100% by water is an island by navigating to Relative Border in Class-Related Features under Feature View as shown in Figure 5.58.

Step 7: Classify image with Class-Related Features and the output should look as shown in Figure 5.59.

In this example we have explored relationship to neighbors.

5.5 TAKING ADVANTAGE OF MORPHOLOGY PARAMETERS

Morphology refers to shape hence, any morphological image processing routine alters or utilizes the shapes and associated shape parameters within an image for analysis. Some of the common morphological operations include:

- Erosion
- Dilation
- Morphologic opening, closing, outlining
- Skeletonization
- Top-hat transform
- Hit-or-miss transform

Erosion reduces the size of objects in relation to their background, and dilation increases the size of objects in relation to their background. These two operations are used to eliminate small object features like noise spikes and ragged edges. Erosion and dilation can be used in a variety of ways, in parallel and in series, to give other transformations including thickening, thinning, skeletonization, and others. Two very important morphological transformations are *opening* and *closing*. Opening generally smooths a contour in an image, eliminating thin protrusions. Closing tends to narrow smooth sections of contours, fusing narrow breaks and long, thin gulfs, eliminating small holes, and filling gaps in contours. Holes are filled in and narrow valleys are closed. Just as with dilation and erosion, opening and closing are dual operations. These morphological filters can be used to eliminate random salt-and-pepper noise present in thematic maps and are useful when minimum mapping units (MMU) are specified for an object.

The background noise is eliminated at the erosion stage under the assumption that all noise components are physically smaller than a predefined structuring element. Erosion will increase the size of the noise components on the object. However, these are eliminated at the closing operation. Morphological operations preserve the main geometric structures of the object. Only features smaller than the structuring element are affected by transformations. All other features at larger scales are not degraded. *Region filling* can be accomplished iteratively using dilations, complementation, and intersections.

The top-hat transformation is a powerful operator, which permits the detection of contrasted objects on nonuniform backgrounds. The top-hat is accompanied by a thresholding operation, in order to create a binary image of the extracted structures. The top-hat block performs top-hat filtering on an intensity or binary image using a predefined neighborhood or structuring element. Top-hat filtering is the equivalent of subtracting the result of performing a morphological opening operation on the input image from the input image itself. This block uses flat structuring elements only.

The hit-and-miss transformation is a general binary morphological operation that can be used to look for particular patterns of foreground and background pixels in an image. It is the basic operation of binary morphology as most of the binary

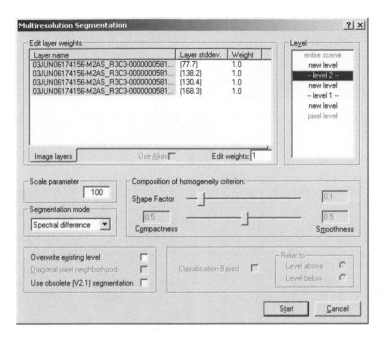

FIGURE 5.60 Using spectral difference function for image segmentation.

morphological operators can be derived from it. As with other binary morphological operators, it takes as input a binary image and a structuring element, and produces another binary image as output. Hit-and-miss operator is the morphological equivalent of template matching, a well-known technique for matching patterns based upon cross-correlation.

eCognition has a function called spectral difference, which is similar to dilation and erosion technique. Let us explore this functionality.

EXAMPLE 5.10: EXPLORING THE SPECTRAL DIFFERENCE IN eCOGNITION

Step 1: Import example13.tif.

Step 2: Segment image at a scale parameter of 10.

Step 3: Segment image at a scale parameter of 100 using spectral difference approach as shown in Figure 5.60.

Notice how well the building edges and smaller details are preserved using the spectral difference approach (Figure 5.61). Figure 5.62 below shows the results of spectral difference segmentation at different scale parameters.

Explore different spectral difference parameters for various feature extraction techniques.

In this example we have learned to perform segmentation based on spectral difference.

FIGURE 5.61 Comparison of image segmentation results with and without spectral difference.

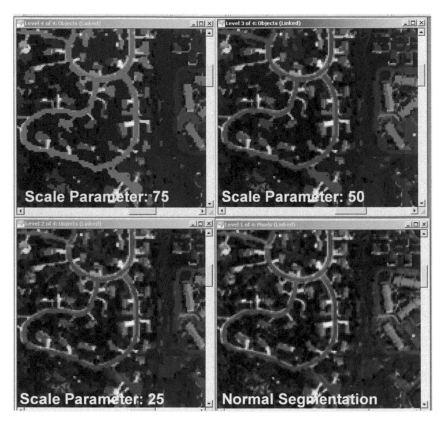

FIGURE 5.62 Image segmentation comparison at different spectral difference scale parameters.

5.6 TAKING ADVANTAGE OF TEXTURE

Texture has been a key component used in the photo interpretation process for identifying different types of forest species and other features. Various texture parameters have been developed to extract texture information from the images. In this section, let us review some of the common texture parameters and associated applications in image analysis.

The textural information is contained in the overall, or average, spatial relationship among gray levels/DNs for a particular image. Texture is indicative of measures of image properties, such as smoothness, coarseness, and regularity. The three principles used to describe the texture context are statistical, structural, and spectral. Statistical techniques use descriptor of spatial relationships and yield characterizations of textures as smooth, coarse, grainy, and other qualitative measures. The spatial relation is considered to be the covariance of pixel values as a function of distance and direction between pixels. Such information can be extracted from an image using gray-tone spatial-dependence matrices or co-occurrence matrices. To analyze a co-occurrence matrix (C) in order to categorize the texture, the statistical parameters as a set of descriptors are computed as follows:

1. Maximum probability — $\max_{ij}(C_{ij})$
2. Second-order inverse element difference moment — $\Sigma_i \Sigma_j \, (i-j)^2 \, C_y$
3. First-order inverse element difference moment — $\Sigma_i \Sigma_j \, C_{ij}/(i-j)^2$
4. Entropy — $\Sigma_i \Sigma_j \, C_{ij} \log(C_{ij})$
5. Uniformity — $\Sigma_i \Sigma_j \, C^2_{ij}$

Gray level co-occurrence matrix (GLCM) and gray level difference vector (GLDV) are some of the common texture measures used in the industry. GLCM texture considers the relation between two neighboring pixels in one offset, as the second-order texture. The gray value relationships in a target are transformed into the C space by a given kernel mask such as 3*3, 5*5, 7*7, and so forth. In the transformation from the image space into the C space, the neighboring pixels in one or some of the eight defined directions can be used; normally, four directions such as 0°, 45°, 90°, and 135° is initially regarded, and its reverse direction can be also taken into account. Hence, GLCM texture measure is dependent upon kernel size and directionality, and known measures such as contrast, entropy, energy, dissimilarity, angular second moment (ASM), and homogeneity are expressed using the aforementioned measures.

GLDV texture measure is the sum of the diagonals of the GLCM texture measure.

GLCM and GLDV texture measures need further interpretation for use in image analysis. Homogeneity is measure for uniformity of C, and if most elements lie on the main diagonal, its value will be large. Dissimilarity measures how different elements of C are from each other. Contrast measures how most elements do not lie on the main diagonal. Entropy is a measure of randomness, and it will be the maximum when the all the elements of C are the same. In case of energy and ASM, they measure extent of pixel pair repetitions and pixel orderliness, respectively.

eCognition has a rich set of texture features as shown in Figure 5.63.

Let us differentiate water from shadows using the texture measures.

EXAMPLE 5.11: USING TEXTURE FOR DISCRIMINATING WATER FROM SHADOWS

Step 1: Import example14.tif.

Step 2: Segment image at a scale parameter of 10.

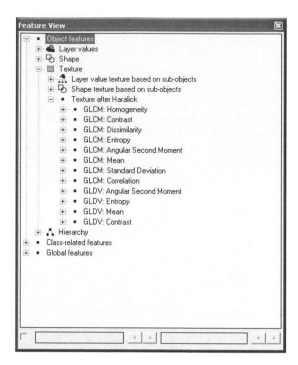

FIGURE 5.63 Texture measures in eCognition.

Step 3: Create a class called Water.

Step 4: Create a new feature called ln[Brightness].

Step 5: Use the following rule sets shown in Figure 5.64 for creating Water classes.

Water bodies tend to have lower standard deviation and homogeneous texture than shadows. Further, blue ratio typically separates low-intensity objects such as water, shadows, and asphalt roads.

Step 6: Classify the project. The output should look similar to the Figure 5.65.

Step 7: Save project.

In this example we have learned to use texture parameters of objects for feature extraction.

5.7   ADDING TEMPORAL DIMENSION

Sun-synchronous land imaging sensors are designed to revisit the same spot on the globe at the same solar time. Several change detection techniques take advantage of the revisit of the satellites for a variety of applications, including natural resources

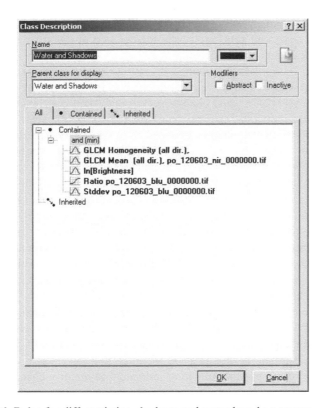

FIGURE 5.64 Rules for differentiating shadows and water based on texture.

FIGURE 5.65 (Color figure follows p. 82.) Thematic map of water derived using texture.

management, urban sprawl monitoring, disaster assessment, encroachments, and others. Images can be acquired by satellites as well as aircrafts during different seasons, lighting conditions, and sensor settings. One of the important aspects of performing change detection is the spatial registration of two or more dates of imagery. Various image processing software, such as Erdas Imagine, PCI,

ER Mapper, and others have the functionality that will allow image-to-image registration as well as second generation orthorectification of one image to another.

Some of the commonly used change detection techniques are: (1) spectral change detection and (2) postthematic change detection. Spectral change detection techniques involve image arithmetic operations, such as image differencing, SAM, cross correlation analysis (CCA), and change vector analysis (CVA). Image differencing can be based on individual spectral bands or derived ratios such as NDVI. This technique captures all the changes between the two scenes. The drawback to this approach is that it also captures changes due to atmospheric changes and does not discriminate between different changes or changes of interest to the end user. Image ratioing is similar to image differencing and suffers from the similar drawbacks. SAM plots the vector of spectral responses of each pixel from two dates. If the vectors are parallel, it is an indicator that the changes are primarily because of the atmospheric differences in the two dates of imagery collection, and deviation of angle vectors from each other is an indicator of change. CCA looks at the covariance of the objects in the two dates. Objects that have high variances from the two dates are categorized as change. CCA has limitations in urban areas. CVA is similar to SAM but uses both direction as well as magnitude of change.

Object-oriented change detection adds a new powerful dimension, parent–child relationship, along with spectral and postthematic techniques for change detection. An example of the power of this technique is change detection of a large bare soil patch now under urban development. If we force the object boundaries from time 1 onto time 2, we can use the following indicators to identify changes:

- Parent–child is now spilt into two or more child objects.
- Child objects that are still bare soil will have similar spectral characteristics as parent objects, and any deviation is an indicator of change.
- Other spectral difference indicators such as CCA and SAM will identify child objects that are different from their parent objects.
- Postthematic classification changes of child versus parent object thematic states.

EXAMPLE 5.12: OBJECTED-ORIENTED CHANGE DETECTION

Step 1: Import dq5013nw.ecw and the associated land use/land cover basemap as thematic layer.

Step 2: Segment the image at a scale parameter of 100 and set the weights of all the image layers to 0 as shown in Figure 5.66.

Click on the Thematic Layers tab to ensure the state says Used. As we want to recreate the land use land cover (LULC) layer, segment the image using a shape-factor of 0. Navigate to Object layers → Thematic Attributes → LULC Code and the display screen should display the LULC boundaries and associated values as shown in Figure 5.67.

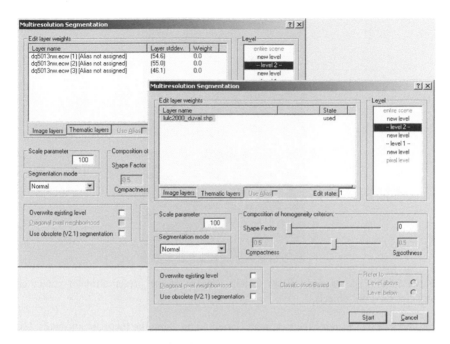

FIGURE 5.66 Assigning different weights to layers for image segmentation.

FIGURE 5.67 Image segmentation using thematic layer.

FIGURE 5.68 (Color figure follows p. 82.) Display of the preservation of parent object boundaries on child objects.

Now we can recreate the base map classification using the thematic attributes. Use the LULCdesc.xls or LULCdesc.txt to see the association between LULC code and class descriptions. For this exercise, let us focus on the polygons selected above, and the LULC code for the selected polygons are 3100 and 4100, respectively. Import the "example5_12.dkb" class hierarchy file already created.

Step 3: Create a new level below the current level making sure the weights for image layers are turned.

Segment the image using a scale parameter of 100 and shapefactor of 0.1. Display object boundaries, and toggle between level 1 and 2 to see the number of objects created at the same scale as shown in Figure 5.68.

Step 4: To display the number of subobjects for basemap objects on Level 2, navigate to Object Features → Hierarchy → Number of Sub-Objects.

You will notice that vegetation objects have a lot of subobjects due to varying texture as well as seasonal differences. In this example, let us focus on urban changes. In the highlighted polygons, you will notice that new construction has taken place in herbaceous and forested lands. Let us explore various eCognition functionalities to identify these changes. Let us investigate the Relationship to Neighbors option as shown in Figure 5.69.

You will notice that all the built-up features have a positive NIR value. Let us explore all the superobject spectral features as shown in Figure 5.70.

Pick one of these sets of features to identify urban changes in the polygons selected, and your output should be similar to Figure 5.71.

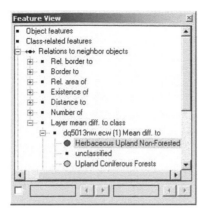

FIGURE 5.69 Using relationship to neighbor feature.

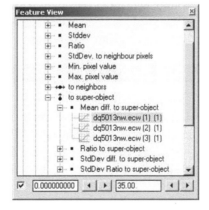

FIGURE 5.70 Exploring relationships to superobject.

FIGURE 5.71 (Color figure follows p. 82.) Object-oriented change detection results.

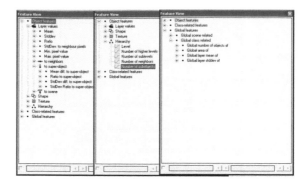

FIGURE 5.72 Change detection parameters within eCognition.

In this example we have learned to explore parent–child object relationships.

Figure 5.72 shows various parameters in eCognition that can be used for change detection. Further, custom relational parameters can be constructed to identify specific changes of interest to the user as shown in Figure 5.73.

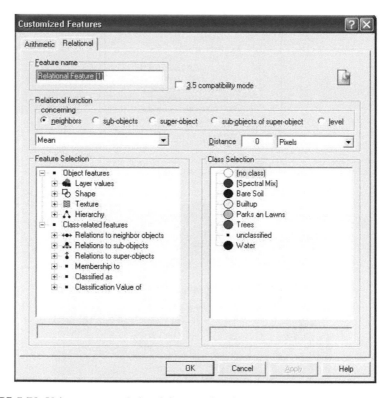

FIGURE 5.73 Using custom relational features for change detection.

6 Advanced Object Image Analysis

This chapter covers advanced image analysis techniques that cover various topics, including how to control image segmentation within eCognition using the parameters available in the software, controlling image segmentation by processing images outside the eCognition environment using image enhancement techniques, such as edge detection and contrast enhancement, and also by importing ancillary information such as parcel data to segment and create objects. This chapter also looks at various image enhancement techniques, such as principal component analysis, and tassel cap transformation, that will help in developing rules bases for extracting features.

6.1 TECHNIQUES TO CONTROL IMAGE SEGMENTATION WITHIN ECOGNITION

As discussed in previous chapters, within eCognition, the image segmentation process can be controlled by the parameter's scale, shape, compactness and smoothness, as well as by importing external thematic layers. We worked on examples where we investigated the effects of scale parameters on object size. Now, let us spend some time understanding the effects of individual parameters on segmentation results. Let us see the effect of shape parameter by segmenting an image at a scale parameter of 25 and using shape parameter values of 0, 0.25, 0.5, and 0.9, respectively, on different levels in eCognition. Although at first glance all the segmented images look alike, there are subtle differences in image segmentation at various shape parameters. If we displayed the number of subobjects from Level 5, you will notice that linear features such as roads are split into more objects increasing shape parameter values as shown in Figure 6.1.

Also, different values of compactness and smoothness factors will result in heterogeneous areas such as parking lots with cars being further split into individual objects as the compactness factor decreases as shown in Figure 6.2. Default parameters of shape are sufficient for most of the applications, whereas image segmentation with higher shape parameters is required for linear features.

6.1.1 USING ANCILLARY GIS LAYERS TO CONTAIN OBJECT BOUNDARIES

One way of constraining image segmentation is by using ancillary data sets. Applications, such as change detection, aggregation of information to parcel/cadastral maps, and creation of land use maps from land cover maps, can use thematic information from base layers to compare objects created on a later date. Using this

FIGURE 6.1 Effects of shape parameter on image segmentation.

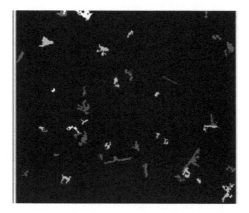

FIGURE 6.2 Effects of varying compactness factors on image segmentation.

approach, let us estimate impervious surfaces on a parcel-by-parcel basis. Thematic layers that can be imported into eCognition include raster images as well as vector files. In this case, we can import parcel boundaries as a shapefile to constrain the segmentation. Example 6.1 shows how to bring in the parcel layer and classify parcels based on the impervious area within each parcel.

EXAMPLE 6.1: CONSTRAINING OBJECT BOUNDARIES WITH PARCELS

Step 1: Import example16.tif.

Step 2: Also import lulc2000_duval.shp as the thematic layer as shown in Figure 6.3.

It is important to note that the thematic layer and the image are properly coregistered for this application. It is possible that the geographic extents of the thematic layer are greater than those of the image layer. To ensure the project extents are the same

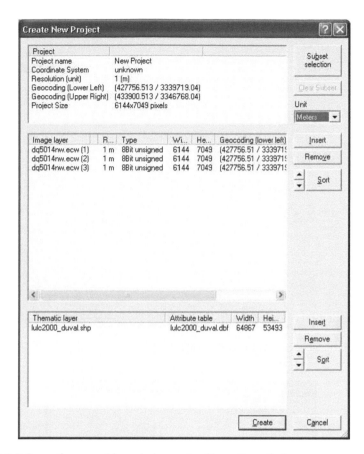

FIGURE 6.3 Importing parcel boundaries vector file as thematic layer.

as that of the image, you can bring in the thematic layer after importing image layers first and using the File → Import Thematic Layer option.

Step 3: Make sure the thematic layer is weighted 1 and the state is used, as shown in Figure 6.4, before segmentation.

Step 4: Segment the layer.

Please note that you can also bring in thematic layers after segmenting the image by using the File → Thematic Layer option. But to make the thematic layer active for segmentation, you need to resegment the level making sure the thematic layer is on.

Step 5: You can check the thematic attribute layers in the Feature View window as shown in Figure 6.5.

Step 6: Create a new layer below using parcel boundaries only.

Step 7: Create a class hierarchy file as shown in Figure 6.6.

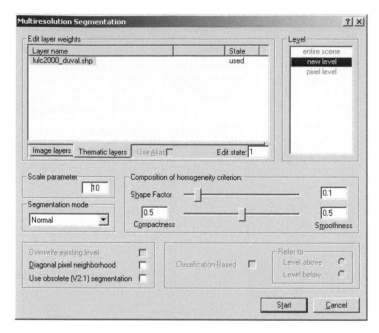

FIGURE 6.4 Image segmentation with thematic layer.

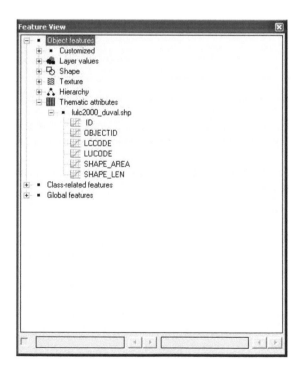

FIGURE 6.5 Accessing thematic attributes in Feature View window.

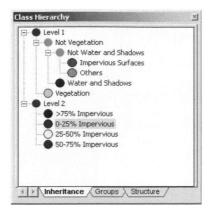

FIGURE 6.6 Class hierarchy for impervious surface aggregation to parcels.

FIGURE 6.7 (Color figure follows p. 82.) Impervious surface classification.

Step 8: Develop rules to create impervious surfaces, based on some of the previous examples.

Classify at Level 1, and the output should look similar to Figure 6.7.

Step 9: Aggregate impervious layers to parcels, and the output should look similar to Figure 6.8.

Step 10: Create vectors of the object boundaries.

Step 11: Export parcel boundaries using the Image Objects option under the Export menu.

FIGURE 6.8 (Color figure follows p. 82.) Impervious surfaces aggregated to parcel vectors.

Step 12: Save project.

In this example, we have learned to:

- Use a shapefile to constrain segmentation
- Aggregate thematic information to imported thematic layer

6.2 TECHNIQUES TO CONTROL IMAGE SEGMENTATION WITHIN ECOGNITION

Let us discuss some of the commonly used techniques for image processing and enhancements that can be created outside eCognition, which will help us constrain segmentation.

6.2.1 SPATIAL FILTERING

Spatial filtering techniques can be used to extract valuable information from an image. These techniques can detect and sharpen boundary discontinuities by exploring the distribution of pixels of varying spectral response across an image. Images typically consist of several dominant spatial frequencies. Fine detail in an image involves a larger number of changes per unit distance than the gross image features. Fourier analysis is the principal technique that can be used for separating an image into its various spatial frequency components. It is possible to emphasize certain groups of frequencies relative to others and recombine the spatial frequencies into an enhanced image. High-pass filters pass high frequencies and emphasize fine detail and edges. Low-pass filters suppress high frequencies and are useful in smoothing an image and may reduce or eliminate salt-and-pepper noise.

The convolution filtering approach is a low-pass filtering approach in which each pixel value is replaced by the average over a window centered on that pixel. This tends to reduce deviations from local averages and smooths the image. The difference between the input image and the low-pass image is the high-pass-filtered output. Edge enhancement filters can highlight abrupt discontinuities, such as rock joints and faults, field boundaries, and street patterns. The Sobel edge enhancement algorithm finds an overabundance of discontinuities and can be used to emphasize the sharp boundaries. Sobel operator S_x and S_y are computed by the convolution of input image and a 3×3 dimension mask, denoted by S_x and S_y. The gradient in x and y directions are computed. The Sobel operators S_x and S_y are shown in Table 6.1.

The gradient magnitude (M) can be calculated from the following equation:

$$M = \text{Sqrt } (S_x^2 + S_y^2) \qquad (6.1)$$

Figure 6.9 shows an image after the Sobel operator has been applied to it. Notice that the streets and highways, and some streams and ridges, are greatly emphasized in the enhanced image. The primary characteristic of a high-pass filter image is that linear features commonly appear as bright lines with dark borders. Details in the water are mostly lost, and much of the image is flat.

Figure 6.10 shows an example of a scene with the low-band-pass operator applied to it. The image is averaged and appears blurry. Now let us look at a high-pass filter image in which the convolution matrix of 7×7 pixels is applied to the image. Figure 6.11 shows an example of an image enhanced using high-pass filter, and you will notice that all the edges appear sharper.

As the number of pixels in the convolution window is increased, the high-frequency components are more sharply defined, as evident from Figure 6.12.

TABLE 6.1
Sobel Operators

$S_x =$			$S_y =$		
−1	0	1	1	2	1
−2	0	2	0	0	0
−1	0	1	−1	−2	−1

FIGURE 6.9 (Color figure follows p. 82.) Image enhancement using a Sobel operator.

FIGURE 6.10 Image enhancement using a low-pass filter.

The next image, Figure 6.13, illustrates the effects of nondirectional edge detectors and a 7 × 7 convolution edge detector.

Based on the application, you can use any of the image enhancement techniques to modify the image for controlling image segmentation in eCognition.

6.2.2 PRINCIPAL COMPONENT ANALYSIS (PCA)

Multiband data sets and images contain redundant information when the bands are adjacent to each other and hence are highly correlated. PCA is a decorrelation

FIGURE 6.11 Image enhancement using a high-pass filter.

FIGURE 6.12 Comparison of various window sizes of high-pass filters.

FIGURE 6.13 Effects of edge detectors on image quality.

procedure that reorganizes by statistical means the DN values from the multispectral bands. It uses eigen analysis for decorrelation. The eigenvector associated with the largest eigenvalue has the same direction as the first principal component. The eigenvector associated with the second largest eigenvalue determines the direction of the second principal component. The sum of the eigenvalues equals the trace of the square matrix, and the maximum number of eigenvectors equals the number of rows (or columns) of this matrix.

A variant of PCA is canonical analysis (CA), which limits the pixels involved to those associated with preidentified features or classes. These pixels are blocked out as training sites, and their spectral values are then processed in the same manner as those of PCA. This selective approach is designed to optimize recognition and location of the same features elsewhere in the scene. Another use of PCA, mainly as a means to improve image enhancement, is known as a decorrelation stretch (DS). This stretch optimizes the assignment of colors that bring out subtle differences not readily distinguished in natural and false color composites. This reduction in inter-band correlations emphasizes small, but often diagnostic, reflectance or emittance variations owing to topography and temperature. The first step is to transform band data into at least the first three principal components (PCs). Each component is rescaled by normalizing the variance of the PC vectors. Then each PC image is stretched, usually following the Gaussian mode. The stretched PC data are then projected back into the original channels, which are enhanced to maximize spectral sensitivity. The difference between the PC color composites and the DS color composites is generally not large. Figure 6.14 shows the four PCA bands.

PCA 2 is a useful layer for the detection of impervious surfaces.

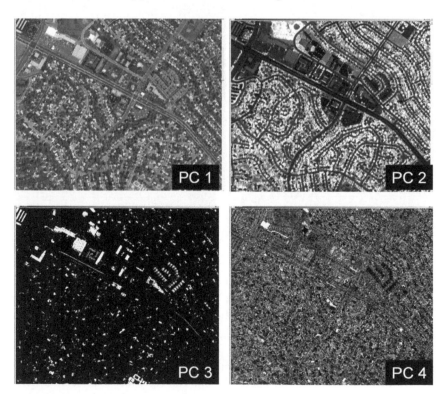

FIGURE 6.14 PCA-transformed bands of an image.

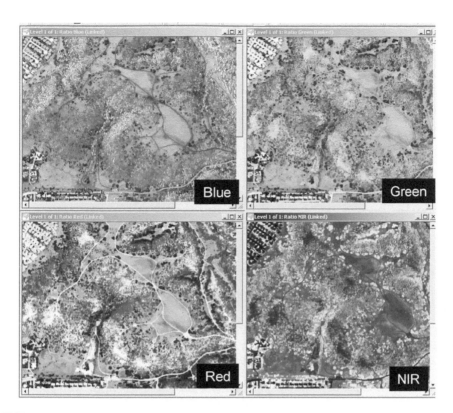

FIGURE 6.15 Ratio of various bands.

6.2.3 RATIOING

Another image manipulation technique that can be used for improving image segmentation is ratioing. Ratioing involves dividing the DN value of each pixel in any one band by the value of another band. One effect of ratioing is to eliminate dark shadows, because these have values near zero in all bands, which tends to produce a truer picture of hilly topography in the sense that the shaded areas are now expressed in tones similar to the sunlight sides.

Figure 6.15 shows the ratios of various ratio bands, where each band is divided by the sum of all the bands.

Notice how the urban features stand out in green and NIR ratios. As we discussed earlier, blue ratio is helpful in identifying water, shadow, and other dark features within the image.

6.2.4 VEGETATION INDICES

Vegetation indices were primarily developed to account for varying atmospheric conditions and eliminate soil background contribution in estimating vegetation responses. There are several popular indices; let us look at some of the popular ones.

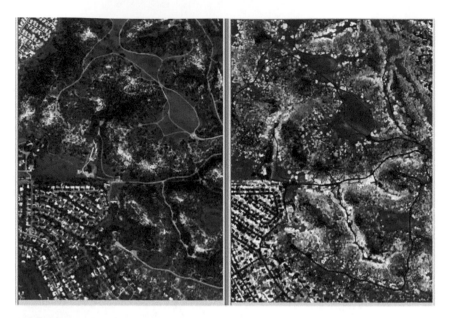

FIGURE 6.16 Ratio Vegetation Index (RVI).

6.2.4.1　Ratio Vegetation Index (RVI)

This index divides one wavelength by the other. NIR increases with increasing canopy, whereas red decreases, and RVI using NIR and red band is represented as follows:

$$RVI = DN_{NIR}/DN_{Red}$$

Typical ranges are a little more than 1 for bare soil, to more than 20 for dense vegetation. Figure 6.16 shows the RVI of a multispectral image.

6.2.4.2　Normalized Difference Vegetation Index (NDVI)

NDVI is the difference in reflectance values divided by the sum of the two reflectance values.

$$NDVI = (\rho_{NIR} - \rho_{Red})/(\rho_{NIR} + \rho_{Red})$$

This index compensates for different amounts of incoming light and produces a number between 0 and 1. The typical range of actual values is about 0.1 for bare soils to 0.9 for dense vegetation. NDVI is usually more sensitive to low levels of vegetative cover, whereas RVI is more sensitive to variations in dense canopies. Figure 6.17 shows the NDVI index of an image.

NDVI-Green replaces the red band with the green band, and Figure 6.18 shows the comparison of NDVI and NDVI-Green images.

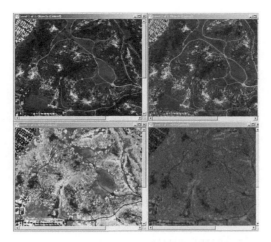

FIGURE 6.17 (Color figure follows p. 82.) Normalized Difference Vegetation Index (NDVI).

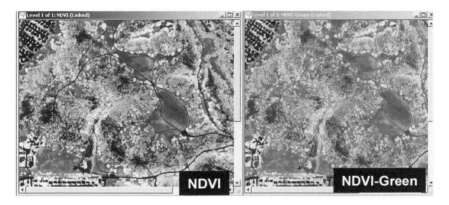

FIGURE 6.18 Comparison of NDVI and NDVI-Green Indices.

6.2.4.3 Soil-Adjusted Vegetation Index (SAVI)

This index resembles the NDVI with some added terms to adjust for different brightness values of background soil.

$$SAVI = ((\rho_{NIR} - \rho_{Red})/(\rho_{NIR} + \rho_{Red} + L)) \times (1 + L)$$

In principle, the term L can vary from 0 to 1 depending on the amount of visible soil. However, 0.5 works as a reasonable approximation for L when the amount of soil in the scene is unknown. TSAVI and TSAVI2 are variations of NDVI to compensate for soil. The Weighted Difference Vegetation Index is also useful for mapping vegetation.

FIGURE 6.19 Results of HIS transformation.

6.2.5 RGB-TO-HIS TRANSFORMATION

Three bands of remote sensor data in RGB color space can be transformed into three bands in HIS color space. This transformation results in bands with correlation to specific features. The equations for this transformation are as follows:

$$I = (R + G + B)/3 \qquad (6.2)$$

$$S = (1 - 3 \, [\min \{R, G, B\}]/(R + G + B) \qquad (6.3)$$

$$H = \cos^{-1} \{[(R - G) + (R - B)])/2 \times [(R - G)^2 + (R - B) (G - B)]^{1/2}\} \quad (6.4)$$

Figure 6.19 shows the hue and intensity components after transformation.

6.2.6 THE TASSEL CAP TRANSFORMATION

The Tassel cap transformation is the conversion of the DNs in various spectral bands into composite values. One of these weighted sums measures roughly the brightness of each pixel in the scene. The other composite values are linear combinations of the values of the separate channels, but some of the weights are negative and others positive. One of these other composite values represents the degree of greenness of the pixels, and another represents the degree of yellowness of vegetation or perhaps the wetness of the soil. Usually, there are just three composite variables.

The tassel cap transformation was designed to model the patterns of crop lands as a function of the life cycle of the crop, using Landsat imagery. A plot of temporal values of NIR and red bands shows that bare soil values typically fall along a line through the origin. During the growing season, as the crop emerges and the canopy starts closing, there is an increase in reflectance in the NIR band due to increasing chlorophyll and a decrease in the red band because of chlorophyll's absorption of red light. At crop maturity, the canopy closes, resulting in peaking of NIR values. As the crop senescence starts, the plants turn yellow. When the crop is harvested, the NIR drops substantially. One component of tassel cap transformation is the weighted sum where the weights are statistically derived and can be characterized as brightness. The second component is perpendicular to the first, and its axis passes through the point of maturity of the plants. The third component corresponds to an axis perpendicular to the first and second axes and passes through the point that represents senescence. The fourth component represents projection onto an axis perpendicular to the other three and usually contains random noise.

The weights used by Kauth and Thomas for the tassel cap transformation of Landsat MSS data are shown in Table 6.2.

Crist and Cicone adapted the tassel cap transformation to the six channels of thematic mapper data. The weights are different, and the third component is taken to represent soil wetness rather than yellowness as in Kauth and Thomas' original formulation. The weights found by Crist and Cicone for the thematic mapper bands are as given in Table 6.3. Table 6.4 and Table 6.5 show the tassel coefficients for IKONOS and QuickBird.

Figure 6.20 shows the tassel cap transformed bands of IKONOS imagery.

Another method that has some merit for consideration is image enhancement to preserve sharp edges. In this example, we will use a contrast stretched image to preserve the sharp boundaries. The edge enhancement was done using ERDAS Imagine Software.

TABLE 6.2
Tassel Cap Coefficients for Landsat MSS

Component	Channel 1	Channel 2	Channel 3	Channel 4
Brightness	0.433	0.632	0.586	0.264
Greenness	−0.290	−0.562	0.600	0.491
Yellowness	−0.829	0.522	−0.039	0.194
Random Noise	0.223	0.012	−0.543	0.810

TABLE 6.3
Tassel Cap Coefficients of TM

Component	Channel 1	Channel 2	Channel 3	Channel 4	Channel 5	Channel 7
Brightness	0.3037	0.2793	0.4343	0.5585	0.5082	0.1863
Greenness	−0.2848	−0.2435	−0.5436	0.7243	0.0840	−0.1800
Wetness	0.1509	0.1793	0.3299	0.3406	−0.7112	−0.4572

TABLE 6.4
Tassel Cap Coefficients of IKONOS

Component	Channel 1	Channel 2	Channel 3	Channel 4
Brightness	0.326	0.509	0.560	0.567
Greenness	−0.311	−0.356	−0.325	0.819
Wetness	−0.612	−0.312	0.722	−0.081

TABLE 6.5
Tassel Cap Coefficients of QuickBird

Component	Channel 1	Channel 2	Channel 3	Channel 4
Brightness	0.319	0.542	0.490	0.604
Greenness	−0.121	−0.331	−0.517	0.780
Wetness	0.652	0.375	−0.639	−0.163

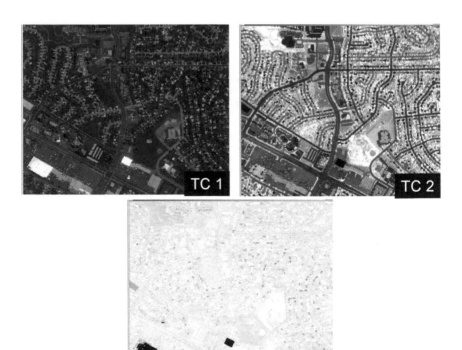

FIGURE 6.20 Tassel cap transformed bands.

EXAMPLE 6.2: CONSTRAINING OBJECT BOUNDARIES USING EDGE-ENHANCED IMAGE

Step 1: Import example6_1.tif.

Step 2: Assign a weight of 1 to the blue channel and 0 to the others. Segment the image using the parameters specified in Figure 6.23.

Step 3: Make sure that the thematic layer is set to On, for segmentation.

Step 4: Segment the image and display object vector outlines, and the image should be like Figure 6.21.

Figure 6.22 shows a comparison of normal segmentation and segmentation using edge-enhanced image.

In this example we have learned to use edge-enhanced raster to constrain segmentation.

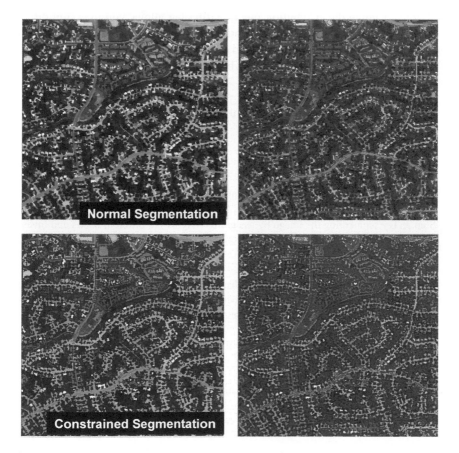

FIGURE 6.21 Image segmentation results with constraints.

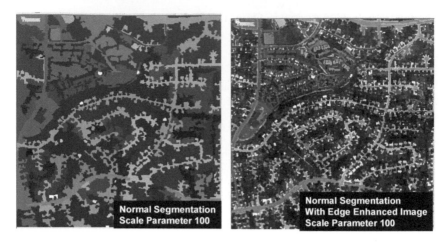

FIGURE 6.22 Edge-enhanced image segmentation comparison.

Let us examine the eCognition functionality that allows you to assign different weights to various input layers. This option is very useful if we have already run some data-mining techniques and figured out which layers are important for feature extraction. Let us run an exercise in which we test object boundaries by giving different weights to layers.

EXAMPLE 6.3: ASSIGNING DIFFERENT WEIGHTS TO LAYERS FOR IMAGE SEGMENTATION

Step 1: Import the four individual bands.

Step 2: Segment the image at a scale parameter of 25.

Step 3: Try segmenting the image by assigning weights* to different bands and investigate the effects of segmentation. Figure 6.24 shows a comparison of segmentation on four individual bands.

You will notice that there are subtle changes in segmentation. Typically, manmade features tend to have high correlation between various spectral bands. Let us investigate the segmentation results on vegetation features.

Step 4: Navigate to vegetation features in the four projects. Figure 6.25 shows a comparison of segmentation results.

Vegetation signatures differ primarily in green and NIR bands. NIR bands capture most of the changes within vegetation. Notice the fine segmentation of vegetation features in NIR-segmented image. You can use feature space optimization to identify

---

* Images must be segmented in different projects as eCognition preserves parent–object relationships.

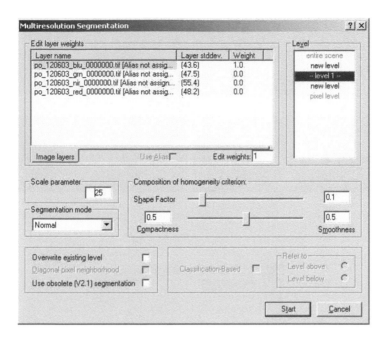

FIGURE 6.23 Assigning different values to image layers for segmentation.

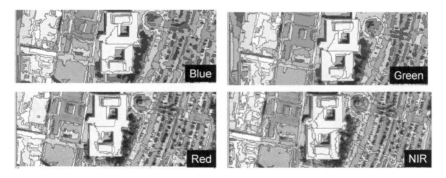

FIGURE 6.24 Segmentation results of individual bands.

the variables that affect the features we are trying to classify. Let us consider an example in which we assign different weights to multiple bands.

In this example we have learned to understand the impact of various bands on segmentation.

EXAMPLE 6.4: FEATURE SPACE OPTIMIZATION

Step 1: Import four individual bands of IKONOS imagery.

Step 2: Segment the image at a scale parameter of 25.

Step 3: Create a class hierarchy with three classes as shown in Figure 6.26.

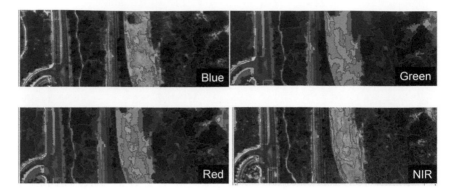

FIGURE 6.25 Comparison of segmentation results of vegetation features for individual bands.

FIGURE 6.26 Class hierarchy for segmentation with different weights.

Step 4: Select samples for each class.

Step 5: Run Feature Space Optimization using all four bands as shown in Figure 6.27.

Step 5: Make sure the maximum dimension is set to 4 and click on Calculate.

Step 6: Click on Advanced, and the window shown in Figure 6.28 will be displayed.

> You should notice that NIR and green are the optimum two bands for this classifi-cation. Run the optimization using the green band only. You will get a distance of 0.05 if you used the green band alone to separate these classes, whereas using NIR only, the distance of separation of these classes is 0.3 as compared to the optimum distance using green and NIR bands at 0.38. The NIR band needs to be assigned a higher weight as compared to the green band.

In this example we have learned to assign different weights to layers to control segmentation.

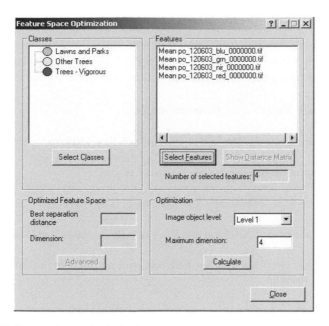

FIGURE 6.27 Feature Space Optimization.

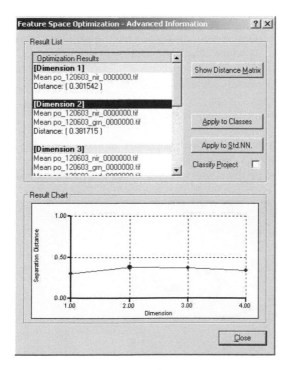

FIGURE 6.28 Feature Space Optimization results.

6.3 MULTISCALE APPROACH FOR IMAGE ANALYSIS

Different features can be classified at different scales for a given application. An example of this would be classifying large water bodies and large bare soil patches at higher scale parameters. This concept will be demonstrated using an example.

EXAMPLE 6.5: MULTILEVEL CLASSIFICATION

Step 1: Import canyon. tif. The band sequence is as follows: NIR, red, green, and SWIR.

Step 2: Segment the image at a scale parameter of 100.

Step 3: Create three classes: Bare Soil [L2], Water [L2], and Others [L2].

Step 4: Use the ratio of NIR band and Stdev of NIR for classifying Water.

Use the rule Not Water and Not Bare Soil by using similarity to classes rule and inverting the rules.

Step 5: Classify and your polygon and output file should look as in Figure 6.29.

Step 6: Merge the objects of Water[L2] and Bare Soil [L2] by performing classification-based segmentation.

Step 7: Let us perform classification-based multiresolution segmentation at a scale parameter of 10 only on the Others[L2] class.

The object boundaries should look as shown in Figure 6.30.

Step 8: Create a class hierarchy as shown in Figure 6.31.

We will use a masking approach to classify various objects. The classes Bare Soil [L1] and Water [L1] use the rule that exploits the relationship of superobjects and existence of respective classes on Level 2. Bare Soil [L1] is created using the inverse

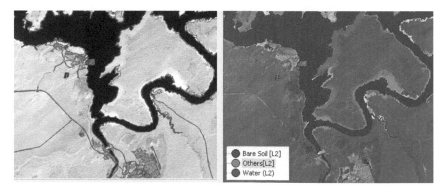

FIGURE 6.29 (Color figure follows p. 82.) Thematic map of soil and water classes.

FIGURE 6.30 (Color figure follows p. 82.) Object boundaries for classification-based multiresolution segmentation.

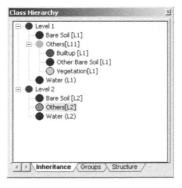

FIGURE 6.31 Class hierarchy for multiresolution segmentation.

similarity rule. Now, we have limited our classification only to Others[L1] class. This contains classes Vegetation, Bare Soil, Water, and Built-up. Use the ratio to identify Water, NDVI for Vegetation, texture parameter GLCM mean in all directions of NIR for Built-up, and other Bare Soil is what is not included in the other three classes.

Step 9: Classify and your output should look as in Figure 6.32.

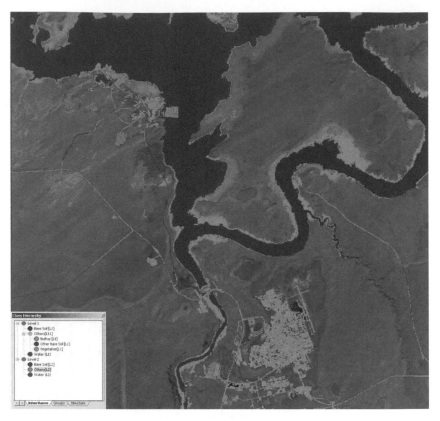

FIGURE 6.32 (Color figure follows p. 82.) Thematic classification based on multiresolution segmentation.

In this example we have learned to:

- Extract various features at different scales
- Use the inverse relationship for class rules
- Use a masking approach for classification

6.4 OBJECTS VERSUS SPATIAL RESOLUTION

Understanding spatial resolution and its influence on object size will determine the features that can be extracted at various pixel sizes. For this exercise, let us use a data set that has been downsampled from 2-ft resolution to different resolutions. The National Imagery Interpretability Rating Scale (NIIRS) we discussed in previous chapters gives you an idea of various features that can be extracted at various resolutions and helps the user select the appropriate resolution for the application at hand. Table 6.6 summarizes the file size for 8-bit uncompressed data sets for 1-band imagery at various resolutions.

TABLE 6.6
Spatial Resolution versus File Size

Spatial Resolution	Number of Pixels/sq km^2	File Size kB
6 in.	44444444	43403
1 ft	11111111	10851
2 ft	2777778	2713
1 m	1000000	977
2.5 m	160000	156
5 m	40000	39
10 m	10000	10
20 m	2500	2
30 m	1111	1

Based on the application at hand, medium-resolution satellites can be a viable alternative for performing some of the thematic classification on large areas. Applications such as watershed modeling, change detection, land use/land cover (LULC) mapping, and others can be accomplished with reasonable accuracy using medium-resolution imagery such as SPOT and Landsat. The following is an example of creating an LULC map on a small subset at multiple resolutions.

EXAMPLE 6.6: LAND USE/LAND COVER CLASSIFICATION AT MULTIPLE RESOLUTIONS

Step 1: Import four bands of IKONOS.

Step 2: Segment the image using default parameters.

Step 3: Import class hierarchy lulc.dkb.

The class hierarchy looks as in Figure 6.33.

Step 4: Assign Aliases for All Bands.

Step 5: Classify the project.

Step 6: Repeat the above steps by importing images at various resolutions.

The combined output should look similar to the Figure 6.34.

In this example we have learned to compare LULC at multiple resolutions.

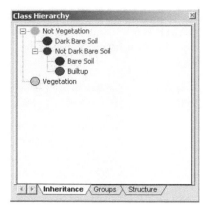

FIGURE 6.33 Class hierarchy for LULC classification at multiple spatial resolutions.

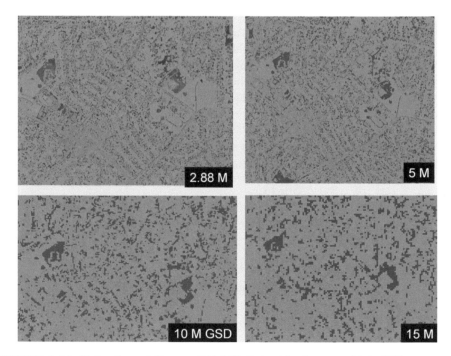

FIGURE 6.34 (Color figure follows p. 82.) LULC classification at multiple GSDs.

6.5 EXPLORING THE PARENT–CHILD OBJECT RELATIONSHIPS

We briefly explored the parent–child relationship in Section 4.2. In this section let us first discuss the importance of having increasing scale parameters for higher levels or vice versa. Let us try an example to segment an image at scale parameter 100 on

FIGURE 6.35 (Color figure follows p. 82.) Parent object boundaries preserved for lower levels of segmentation within eCognition.

Level 1 and scale parameter 25 on Level 2. You will notice that the object boundaries from Level 2 have not changed in Level 1. eCognition preserves the parent–child object relationships as shown in Figure 6.35.

Also, note that while using the Merging Objects option, the option will not merge objects of the same classification contiguous to each other unless the objects are merged at all the levels above to ensure parent–child object relationship. Let us do an example where we use Level 2 as a masking plane and derive impervious surfaces on Level 1.

Example 6.7: Using a Masking Approach to Thematic Classification

Step 1: Import the tif file QuickBird.

Step 2: Segment the image at a scale parameter of 100.

Step 3: Segment the image at scale parameter 25 one level below.

We will create a mask layer on Level 2.

Step 4: Create a class hierarchy as in Figure 6.36.

Step 5: Now, using the Manual Classification option, create masks that are not impervious surfaces on Level 2, as shown in Figure 6.37.

Step 7: Now, let us develop a rule for impervious areas that says any object classified as mask in Level 2 is not an impervious area. Further, use NDVI and NIR ratios to develop class rule sets for impervious surfaces. Select appropriate rule sets for Vegetation and Bare Soil.

Step 8: Classify.

Step 9: Save project.

FIGURE 6.36 Class hierarchy file.

FIGURE 6.37 Mask on Level 2.

This approach allows you to eliminate all the areas that do not need to be included in the processing. If you merge all the mask objects in Level 2 by using classification-based segmentation, you can decrease the number of objects created on Level 1, saving processing time. Also, you can bring in ancillary data on Level 2 to complement or replace the manual mask creation process.

In this example we have learned to use manual classification and associated editing tools.

Other parent–child properties can be useful in thematic classification. As discussed in the change detection example, spectral difference between child and parent–child objects is an important aspect that can be used in updates of thematic maps. Other spectral relationships between parent and child objects, including standard deviation differences and ratio differences, are helpful in feature extraction. Another simple but powerful application is the creation of masks using parent objects and limiting the thematic classification only to the child objects of interest. This approach can also be used to segment child objects at a finer segmentation scale only on parent objects of interest. Texture differences between parent and child objects can be leveraged to identify outliers in child objects that deviate from parent objects. This property can be exploited in updating thematic maps.

6.6 USING SEMANTIC RELATIONSHIPS

One of the powerful features within eCognition is semantic grouping. We will be using the Groups tab in the Class Hierarchy window for this. This function can be used to group classes that belong together. An example of semantic grouping is estimated cloud cover within a given scene. If we were to classify clouds and cloud shadows, we can semantically group these two groups to compute percentage cloud cover in the scene. All the rules that apply to individual classes can be extended to semantic groups, such as distance measures, within eCognition. The creation of semantic groups is now examined:

EXAMPLE **6.8**: SEMANTIC GROUPING

Step 1: Open example4_3.dps.

Step 2: Create two additional classes, the Pervious and Impervious Group.

Step 3: Navigate to Groups tab and drag the classes that belong to the Pervious and Impervious Group, respectively, as shown in Figure 6.38.

Step 4: Use the left and right arrow icons to display the LULC map and the pervious/impervious map created using semantic grouping. The maps should look as shown in Figure 6.39.

Now let us perform some spatial operations on the semantic groups. Click on Relative Border to Pervious Group, and you will see that the relationships between objects and semantic groups can now be used for image analysis.

Let us consider another example that involves computing percentage cloud cover in an image using the semantic relationship.

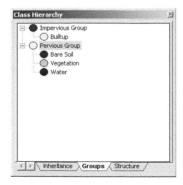

FIGURE 6.38 Semantic grouping.

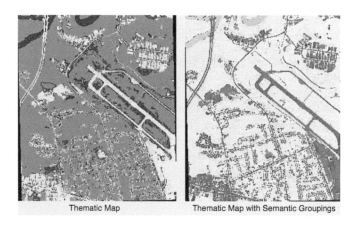

FIGURE 6.39 (Color figure follows p. 82.) Thematic map with semantic grouping.

EXAMPLE 6.9: SEMANTIC GROUPING

Step 1: Import example23.tif.

Step 2: Segment image at a scale parameter of 100.

Step 3: Create a class hierarchy Clouds and Cloud Shadows.

Step 4: Use the following rule set to classify Clouds and Cloud Shadows as shown in Figure 6.40.

Step 5: Merge the objects for Clouds and Cloud Shadows.

Step 6: Create a class called Cloud Group.

Step 7: Drag the classes Cloud and Cloud Shadows under Cloud Group.

FIGURE 6.40 Class hierarchy rules for cloud detection.

Step 8: Display the Cloud group, and it should look similar to Figure 6.41.

Step 9: Let us investigate the global relationships by navigating to Global Features in Feature View as shown in Figure 6.42.

Step 10: Save project.

In this example we have learned to:

• Explore semantic relationships
• Investigate global relationships

6.7 TAKING ADVANTAGE OF ANCILLARY DATA

Ancillary data sets such as the elevation data, existing planimetric data, and other GIS data sets should be used as needed to assist in image analysis. Let us discuss a few scenarios in which we can use ancillary data sets. An example can be of vegetation classification within a watershed for estimation of carbon modeling.

FIGURE 6.41 (Color figure follows p. 82.) Semantic grouping of clouds and cloud shadows.

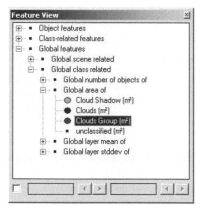

FIGURE 6.42 Global related features.

Ancillary data sets of watershed boundary along with stream vectors, created outside eCognition software, can be brought in as thematic layers for image analysis. The watershed boundary can be used as a mask to restrict the analysis within the boundary.

Another example would be the forest species classification, where ancillary data sets that represent the knowledge of phenomenology, geographic distribution, and other information can be used to classify certain forest species. Similarly, elevation data from Lidar can be effectively combined with multispectral data to extract buildings and road networks.

7 Accuracy Assessment

Accuracy assessment is an integral part of information extraction and requires a priori planning on the approach to measure the accuracy of the information derived from multispectral imagery. Remotely sensed information products are increasingly used in decision-making applications, such as computing taxes based on the area of impervious surfaces in a given parcel or estimating forest fire potential based on tree count, and others. Accuracy assessment is critical to ensuring the accuracy of the information used in the decision process. Often, ground truth or collateral information is required for statistical analysis. There are several protocols that have been developed to collect ground truth using global positioning systems (GPS). These protocols differ for linear features as compared to features such as land cover, and the user should employ an appropriate technique based on the information extracted.

Further, selecting a statistically significant number of samples is an important aspect to consider for assessing the accuracy of a thematic map. Once the samples are selected, a gamut of statistical techniques is available for measuring the final accuracy of the information extracted. The following sections will give an overview of various sampling techniques and accuracy assessment measures commonly used in the remote sensing/mapping industry.

7.1 SAMPLE SELECTION

An effective sampling method requires some means for establishing sample sizes and appraising the sample results mathematically. A sample set should have a behavior that is measurable in terms of the rules of the theory of probability. Using statistical sampling one can, within a stipulated degree of confidence, conclude that the number of errors in the sample applies proportionately to the unsampled portion of the universe as well. Advantages of statistical sampling include the following:

- Objective and defensible techniques can be used, and the approach is not subject to bias.
- Based on the desired level of confidence, one can plan for the size of the maximum sample needed, sample size, and justification for expense or time spent prior to the start of the project.
- Allows for estimating the degree of risk, based on the number and spatial distribution of the sample set selected, including the possibility that the set may not be representative of the entire population.

- Statistical sampling provides a means of projecting test results within known limits of reliability.
- Use established techniques to optimize the use of samples such as bootstrap and jackknife statistical sampling approaches.

A minimum mapping unit is often associated with measuring the accuracy of nonlinear features such as land use/land cover classes. The size of the minimum mapping unit depends on a variety of factors, including the final application of the derived information product, the spatial resolution of the pixels in the image, type of features extracted, and other factors. Minimum mapping units typically have window sizes of 3×3 pixels or greater, or can be represented in units such as $^{1}/_{2}$ acre to 1 acre.

7.2 SAMPLING TECHNIQUES

The population distribution should be used to determine the sample selection technique. The specific plan and selection technique used should be precisely documented. The sampling techniques most often used are as follows:

Random sampling: Each item in the population has an equal chance of being included in the sample. This is the most common method of sampling.

Interval sampling: The sample items are selected from within the universe in such a way that there is a uniform interval between each sample item selected after a random start.

Stratified sampling: The items in the population are segregated into two or more classes or strata. Each stratum is then sampled independently. The results for the several strata may be combined to give an overall figure for the universe or may be considered separately, depending on the circumstances of the test.

Cluster sampling: The universe is formed into groups or clusters of items. Then, the items within the selected clusters may be sampled or examined in their entirety.

Multistage samples: Involves sampling at several levels. As an example, the user takes a sample from several locations and then, in turn, takes another sample from within the sampled items.

7.3 GROUND TRUTH COLLECTION

Ground truth or collateral information collection is an important step in validating the accuracy of a thematic map. While sending personnel to the field, it is important to establish protocols for data collection and to ensure consistency. Depending on the type of the feature and its size and accessibility, the protocols will differ. For extracting collateral information from the imagery itself, most of the commercial image processing software have functionality to select samples using random as well as stratified random sampling. It is important to also collect a few ground control points (GCPs) along with the ground truth that will allow tying the image precisely to the ground coordinates.

7.4 ACCURACY ASSESSMENT MEASURES

In this section, let us discuss some of the common measures used for accuracy assessment. The following is a list of measures used for measuring the accuracy of thematic maps derived from multispectral imagery:

- Error (confusion) matrix
- Error of comission
- Error of ommission
- Producer's accuracy
- Consumer's accuracy
- Overall accuracy
- Kappa coefficient

Error matrix: Error matrix is a matrix that shows the comparison of thematic classes with reference or sample data. Bias occurs when a classification estimate is optimistic or conservative. The three significant sources of conservative bias include (1) errors in reference data, (2) positional errors, and (3) minimum mapping unit. Similarly, some of the significant sources of optimistic bias include (1) use of training data for accuracy assessment, (2) restriction of reference data sampling to homogeneous areas, and (3) sampling of reference data not independent of training data. The magnitude and direction of bias in classification accuracy estimates depend on the methods used for classification and reference data sampling. Reporting an error matrix and associated classification accuracy estimates is not sufficient and should be accompanied by a detailed description of methods used, to assess the potential for bias in the classification accuracy estimates.

Error of commission: Error of commission is a measure of the ability to discriminate within a class, and occurs when the classifier incorrectly commits pixels of the class being sought to other classes.

Error of omission: An error of omission measures between class discrimination and results, when one class on the ground is misidentified as another class by the observing sensor or the classifier.

Producer's accuracy: The producer of the classification wants to predict each ground truth point correctly. Producer's accuracy is computed by looking at the reference data for a class and determining the percentage of correct predictions for these samples.

Consumer's accuracy: Consumer's or user's accuracy is computed by looking at the predictions produced for a class and determining the percentage of correct predictions.

Overall accuracy: Overall accuracy is computed by dividing the total correct (sum of the major diagonal) by the total count in the error matrix.

Kappa coefficient: Kappa coefficient is a measure of the interrater agreement. When two binary variables are attempts by two metrics to measure the same thing, you can use Kappa as a measure of agreement between the two metrics.

TABLE 7.1

Interpretation of Kappa Values

Kappa Value	Interpretation
Below 0.00	Poor
0.00–0.20	Slight
0.21–0.40	Fair
0.41–0.60	Moderate
0.61–0.80	Substantial
0.81–1.00	Almost perfect

$$k = P_v - P_e/1 - P_e \qquad (7.1)$$

where $P_v = \Sigma_i P_{ii}$ and $P_e = \Sigma_i P_i P_i$.

If the two response variables are viewed as two independent ratings of the n subjects, the kappa coefficient equals +1 when there is complete agreement of the raters. When the observed agreement exceeds chance agreement, kappa is positive, with its magnitude reflecting the strength of agreement. Although this is unusual in practice, kappa is negative when the observed agreement is less than the chance agreement. The minimum value of kappa is between 1 and 0, depending on the marginal. Table 7.1 shows the general interpretation rules for thematic accuracy assessment.

eCognition has built-in functionality for performing accuracy assessment. Let us perform accuracy assessment on the land use/land cover product. eCognition provides accuracy assessment methods serving different purposes. These tools produce statistical and graphical outputs that can be used to check the quality of the classification results. The tables from the statistical results can be saved in comma-separated ASCII text files. The graphical results can be exported in their current view settings as several raster files. In addition, eCognition provides a powerful statistics tool with basic GIS functionality. These tables can also be saved as ASCII text files with comma-separated columns.

eCognition offers the following four methods:

Classification stability: This feature allows us to explore the differences in degrees of membership between the best and the second-best class assignments of each object that can give evidence about the ambiguity of an object's classification. The graphical output can be displayed by selecting Classification Stability in the View Settings dialog box. The value is displayed for each image object in a range from dark green (1.0, crisp) to red (0.0, absolutely ambiguous).

Best classification result: This tool generates a graphical and statistical output of the best classification results for the image objects of a selected level.

Because of eCognition's fuzzy classification concept, an image object has memberships in more than one class. The classification with the highest assignment value is taken as the best classification result.

Error matrix based on TTA mask: This method uses test areas as a reference for classification quality. Test areas can be generated outside eCognition and imported into an eCognition project by means of a TTA mask to compare the classification with ground truth based on pixels.

Error matrix based on samples: This tool is similar to that mentioned in the previous paragraph but considers samples derived from manual sample inputs.

EXAMPLE 7.1: PERFORMING ACCURACY ASSESSMENT

Step 1: Import example24.tif.

Step 2: Segment the image at a scale parameter of 25.

Step 3: Create another level at a scale parameter of 25.

We will use one level to perform classification and the other to perform accuracy assessment.

Step 4: Create a class hierarchy as shown in Figure 7.1.

Step 5: On Level 1, select samples for all the classes and perform supervised classification. The thematic map should look similar to Figure 7.2.

Step 6: Using the same samples used for supervised classification, let us perform accuracy assessment.

Step 7: Navigate to Tools → Accuracy Assessment.

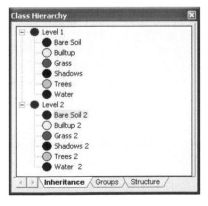

FIGURE 7.1 Class hierarchy for accuracy assessment project.

FIGURE 7.2 (Color insert follows p. 82.) Thematic map for accuracy assessment project.

FIGURE 7.3 (Color insert follows p. 82.) Classification stability display.

Step 8: Let us look at the option classification stability based on the samples, and the display screen should look as in Figure 7.3.

The green values show stable classes, whereas shades of red show classes that are not stable. You will notice that there is confusion between bare soil, built-up, and areas with patches of vegetation. This function can be used for quality assessment to identify objects that need further investigation.

Step 9: Now let us investigate the option, Best Classification Result. This window should look as in Figure 7.4.

The green color shows the classes that have membership degrees close to 1, and the red color shows the ones that are unstable classes and have low membership values. It is important to note that an object can have high membership values for more than one class such as shadows and water bodies.

FIGURE 7.4 (Color insert follows p. 82.) Best Classification window.

FIGURE 7.5 TTA mask.

In practice, it is not recommended that the training sample set be used to perform accuracy assessment. When training samples are limited, there are statistical sampling techniques such as jackknife and bootstrap approaches that allow optimum use of training sets. Let us create an independent set of samples for accuracy assessment on Level 2.

Step 10: Navigate to Level 2.

Step 11: Use the rule of dependency on subobject classification to classify Level 2.

Step12: You can select samples for each of the classes on Level 2 and perform accuracy assessment. eCognition also allows us to bring in samples created either from a different eCognition project or other software such as ERDAS Imagine. Let us bring in some samples that I have already created. To accomplish this, navigate to Samples → Load TTA Mask. The TTA Mask option allows eCognition users to export and import samples from eCognition. The TTA mask should appear in Figure 7.5.

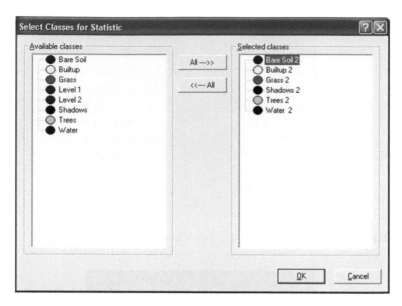

FIGURE 7.6 Selecting classes for accuracy assessment.

User \ Referenc...	Shadows 2	Bare Soil 2	Water 2	Builtup 2	Trees 2
Confusion Matrix					
Shadows 2	0	0	0	0	0
Bare Soil 2	303	4821	0	0	0
Water 2	4078	0	27511	0	264
Builtup 2	0	722	396	13003	0
Trees 2	0	0	0	0	3414
Grass 2	0	1316	0	0	0
unclassified	0	0	0	0	0
Sum	4381	6859	27907	13003	3678
Accuracy					
Producer	0	0.7029	0.9858	1	0.9282
User	undefined	0.8911	0.8637	0.9208	1
Hellden	0	0.7859	0.9207	0.9588	0.9628
Short	0	0.6473	0.853	0.9208	0.9282
KIA Per Class	0	0.6732	0.9696	1	0.9239
Totals					
Overall Accuracy 0.8765					

FIGURE 7.7 Accuracy assessment table.

Step 13: Perform accuracy assessment based on TTA mask. Select classes on Level 2 only as shown in Figure 7.6.

Perform accuracy assessment, and the report should look similar to Figure 7.7.

Step 14: eCognition allows you to export the report into an ASCII format. This report can be supplied to the end user along with the thematic map.

In this example we have learned to:

- Perform accuracy assessment
- Bring in samples through the TTA mask
- Investigate various accuracy measures in eCognition

References

Albuquerque, T., Barata, T., Pina, P., and Jardim, S., 2003, Classification through Mathematical Morphology and Geostatistical approaches using Landsat TM images, http://www. gruponahise.com/simposio/papers%20pdf/34%20Teresa%20Albuquerque.pdf.

Baatz, M., Benz, U., Dehghani, S., Heynen, M., Höltje, A., Hofmann, P., Ligenfelder, I., Mimler, M., Sohlbach, M., Weber, M., and Willhauck, G., 2000, eCognition User Guide 4.

CCS[3], Los Alamos Laboratory, 2003, Law's Texture Measures, http://public.lanl.gov/kelly/ Notebook/laws.shtml.

Chiu, W.Y. and Couloigner, I., 2004, Evaluation of Incorporating Texture into Wetland Mapping from Multi-Spectral Images, http://las.physik.uni-oldenburg.de/eProceedings/ vol03_3/03_3_chiu1.pdf.

Corner, B.R., Narayan, R., and Reichenbach, S., 1999, Principal Component Analysis of Remote Sensing Imagery: Effects of Additive and Multiplicative Noise, http://doppler. unl.edu/~bcorner/spie99.pdf.

Crist, E.P. and Cicone, R.C., 1984, Application of the Tasseled Cap concept to simulated Thematic Mapper data, *Photogrammetric Engineering and Remote Sensing*, Vol. 50, No. 3, 343–352, March 1984.

Federation of American Scientists (FAS), 1998, National Imagery Interpretability Rating Scales, http://www.fas.org/irp/imint/niirs.htm.

Foley, J.D., van Dam, A., Feiner, S.K., and Hughes, J.F., 1990, *Computer Graphics: Principles and Practice,* 2nd ed., Boston: Addison-Wesley Publishing Company.

Fraser, N., 2001, Neil's Neural Nets, http://vv.carleton.ca/~neil/neural/.

Gonzales, R.C. and Woos, R.E., *Digital Image Processing*, Boston: Addison-Wesley Publishing Company, 1992.

Green, B., 2002, Canny Edge Detection Tutorial, http://www.pages.drexel.edu/~weg22/ can_tut.html.

Hall-Beyer, M., 2006, The GLCM Tutorial Home Page, http://www.fp.ucalgary.ca/mhallbey/ texture_calculations.htm.

Heath, M., Sarkar, S., Sanocki, T., and Bowyer, K.W., A Robust Visual Method for Assessing the Relative Performance of Edge-Detection Algorithms IEEE, *Transactions on Pattern Analysis and Machine Intelligence*, Vol. 19, No. 12, 1338–1359, December 1997.

Jakomulska, A. and Clarke, K.C., 2001, Variogram-Derived Measures of Textural Image Classification, http://www.geog.ucsb.edu/~kclarke/Papers/Jakamulska.pdf.

Jensen, J.R., 2000, *Remote Sensing of the Environment: An Earth Resource Perspective*, Upper Saddle River, NJ: Prentice Hall.

Jensen, J.R., 2005, *Introductory Digital Image Processing*, 3rd ed., Upper Saddle River, NJ: Prentice Hall.

Kauth, R.J. and Thomas, G.S., 1976, The Tasseled Cap — A Graphic Description of the Spectral-Temporal Development of Agricultural Crops as Seen by Landsat, *Proceedings of the Symposium on Machine Processing of Remotely Sensed Data*, Purdue University, West Lafayette, IN, pp. 4B41–4B51.

LARS, Purdue University, 2006, Multispectral/Hyperspectral Image Information, http://www. lars.purdue.edu/home/image_data/image_data.htm.

Lee, K., Jeon, S.H., and Kwon, B.D., 2004, Urban Feature Characterization using High-Resolution Satellite Imagery: Texture Analysis Approach, http://www.gisdevelopment. net/technology/ip/ma04228pf.htm.

Lillesand, T.M., and Kiefer, R.W., *Remote Sensing and Image Interpretation*, 3rd ed., New York: Wiley publishing company.

Lucieer, A., Fisher, P., and Stein, A., 2003, Fuzzy Object Identification Using Texture Based Segmentation of High-Resolution DEM and Remote Sensing Imagery of a Coastal Area in England, http://www.itc.nl/personal/lucieer/downloads/lucieer_sdq2003_abstract.pdf.

Math Works, 2006, Video and Image Processing Blockset, http://www.mathworks.com/access/helpdesk/help/toolbox/vipblks/f6010dfi1.html.

Middlesex University of Computing Science, 2006, Segmentation, http://www.cs.mdx.ac.uk/staffpages/peter/com3325/p3.pps.

Min, J., Powell, M.W., and Boyer, K.W., 2000, Range Image Segmentation, http://marathon.csee.usf.edu/range/seg-comp/SegComp.html.

NCGIA, University of California, Santa Barbara, Road centerlines from hyper-spectral data, http://www.ncgia.ucsb.edu/ncrst/resources/easyread/HyperCenterlines/first.html.

Short, N.M., Sr., 2006, Remote Sensing Tutorial, http://rst.gsfc.nasa.gov/Front/tofc.html.

Park, M., Jin, J.S., Wilson, L.S, 2000, Hierarchical Indexing Images Using Weighted Low Dimensional Texture Features, http://www.cipprs.org/vi2002/pdf/s1-6.pdf.

Proakis, J.G., *Digital Communications*, 2nd ed., New York: McGraw-Hill.

Quantitative Imaging Group, The Netherlands, 2006, Image Processing Fundamentals, http://www.ph.tn.tudelft.nl/Courses/FIP/noframes/fip-Morpholo.html#Heading99.

Ray, T.W., 1994, A FAQ on Vegetation in Remote Sensing, http://www.yale.org/ceo/Documentation/rsvegfaq.html.

Robinson, J.W., Principal Component Analysis: A Background, http://rst.gsfc.nasa.gov/AppC/C1.html.

Silvia Merinao de Miguel, 1999, Geostatistics and Remote Sensing: Forestry Applications, http://www.geog.ucl.ac.uk/~smerino/ and http://www.lib.unb.ca/Texts/JFE/bin/get2.cgi?directory=July97/&filename=murphy.html.

Skirvin, S.M. et al., 2000, An Accuracy Assessment of 1997 Landsat Thematic Mapper, Derived Land Cover for the Upper San Pedro Watershed (US/Mexico), http://www.epa.gov/esd/land-sci/pdf/037leb01.pdf.

Stoney, W.E., 2005, Guide to Land Imaging Satellites, http://www.asprs.org/news/satellites/ASPRS_DATABASE_121905.pdf.

Suzuki J. and Shibasaki, R., 1997, Development of Land Cover Classification Method Using Noaa Avhrr, Landsat TM and DEM Images, http://www.gisdevelopment.net/aars/acrs/1997/ps3/ps3002pf.htm.

Verbyla, D., How To Lie With An Error Matrix, http://nrm.salrm.uaf.edu/~dverbyla/online/errormatrix.html.

Virtual Laboratory for GIS and Remote Sensing, 2000, Background on Spectral Signatures, http://geog.hkbu.edu.hk/virtuallabs/rs/env_backgr_refl.htm.

Visual Numerics, 2005, PV-Wave Family, http://www.vni.com.tw/tw/products/wave/ip_flyer.html.

Watkins, T., Principal Component Analysis in Remote Sensing, http://www2.sjsu.edu/faculty/watkins/princmp.htm.

Weier J. and Herring, D., Measuring Vegetation, http://earthobservatory.nasa.gov/Library/MeasuringVegetation.

Wikipedia, 2006, Principal Component Analysis, http://en.wikipedia.org/wiki/Principal_components_analysis.

Yarbrough, L.D., 2005, QuickBird 2 Tassel Cap Transformation Coefficients: A comparison of Derivative Methods, Pecora 16, Global Priorities in Land Remote Sensing.

Yuen, P., 1998, Tasseled Cap Transform Breathing New Life into Old Applications, http://www.imagingnotes.com/old/novdec98/novstory6.htm.

Related Titles

Laser Remote Sensing
Takashi Fujii, Central Research Institute of Electric Power, Japan
ISBN: 0-8247-4256-7

Introduction to Microwave Remote Sensing
Iain Woodhouse, University of Edinburgh, Scotland
ISBN: 0-415-27123-1

Remote Sensing of Snow and Ice
W. Gareth Reese, Cambridge University, UK
ISBN: 0-415-29831-8

Index

T - #0440 - 071024 - C24 - 234/156/9 - PB - 9780367446246 - Gloss Lamination